AF589463

Indian Botanic Gardens

The Editors

Dr R.K.Roy, is a renowned horticulturist in India. He has worked in reputed research institute like CSIR-National Botanical Research Institute (NBRI), Lucknow and Agricultural and Horticultural Society of India, Kolkata in various capacities. Dr. Roy, was former Sr. Principal Scientist and Head, Botanic Garden, Floriculture and Landscaping Division, CSIR-NBRI (1989-2016). Prior to that, Dr. Roy was Assistant Secretary and Horticulturist at The Agricultural–Horticultural Society of India, Alipore, Kolkata (1985-89). His main field of specialization is Ornamental Horticulture, Floriculture and Landscaping. During the tenure of his service spanning over 35 years, Dr. Roy has been deeply involved in the R&D work on Floriculture and Landscaping.

Dr Roy is member of various professional and academic societies of India and abroad. He is Sr. Vice President of Bougainvillea Society of India and Former Vice-Chairman, Commission Landscape and Urban Horticulture, International Society for Horticultural Sciences (ISHS), Belgium.

He was also awarded with international fellowship/scholarship *viz*. **'RHS's Financial Award'** by Royal Horticultural Society, London, United Kingdom in 1998; **'Commonwealth Science Council Fellowship'**, Commonwealth Science Council, United Kingdom in 2003; **'Coke Trust Award'** by Royal Horticultural Society, United Kingdom. Moreover, Dr. Roy had international exposure by way of visiting many countries and leading botanical/ornamental gardens in United Kingdom, France, Poland, Switzerland, Germany, The Netherland *etc.*

He has written many books *viz.* 'Ornamental Annuals', 'House Plants', 'Kitchen Garden', 'Home Garden', 'Gladiolus', 'Gerbera', 'Marigold', 'Fundamentals of Garden Designing' and 'Ornamental Trees of India' *etc.*

Dr J.S. Khuraijam is a plant conservationist and cycad biologist at CSIR-National Botanical Research Institute, Lucknow. He has 12 years of research experience and his major contribution has been in the field of cycad taxonomy and conservation of threatened species. Besides curation of living collections at CSIR-NBRI Botanic Garden, he has been actively involved in the taxonomic studies of conifers, orchids and conservation of wild ornamental plants. He has published more than 60 papers on plant taxonomy and biodiversity conservation and has edited two books. Dr Khuraijam did his Ph.D in Botany (2014) from GGS Indraprastha University, New Delhi. He is an active member of IUCN SSC Cycad Specialist Group.

Indian Botanic Gardens

R.K. Roy

J.S. Khuraijam

2020

Daya Publishing House®

A Division of

Astral International Pvt. Ltd.

New Delhi – 110 002

ISBN 9789390371372 (Int. Edition)

Published by : Daya Publishing House®
A Division of
Astral International Pvt. Ltd.
– ISO 9001:2015 Certified Company –
4736/23, Ansari Road, Darya Ganj
New Delhi-110 002
Ph. 011-43549197, 23278134
E-mail: info@astralint.com
Website: www.astralint.com

Preface

One fourth of all the plant species in the world are at risk of being endangered or going extinct. In India, more than 2000 plant species are already threatened and need urgent conservation efforts. Natural calamities triggered by human activities have resulted in the rapid increase of extinction of species and destruction of their natural habitats. Extinction of wild species posed an adverse effect on the ecological balance. Climate change, deforestation, soil erosion, flood, landslides have made the ecosystem vulnerable to permanent reduction in carrying capacity for all organisms. The survival of ecosystems depends on the biodiversity, habitats and the interactions among the species. Understanding the ecosystem types, habitats, associated species and breeding play an important role in the long term conservation of biodiversity.

Ex-situ conservation of plant diversity in botanic gardens and arboretum is an effective conservation measure adopted worldwide for the conservation of economically important, ornamental, rare and threatened plant species. Through the sum of knowledge and expertise that they have accumulated, botanic gardens are leaders in scientific research on both wild and cultivated plants and conservation. Botanical gardens are a unique environment to educate and create public awareness about the importance of biodiversity, the threats it currently faces and make them realize that nature conservation is everyone's job. This is why it is so important for gardens to maintain interpretation programs, host school and college groups, present exhibitions and organise flower shows to promote nature conservation and healthy environment.

Activities of botanic gardens depend on the climatic zones and expertise in the botanic gardens. The collaboration among the botanic gardens and also with other organizations such as forest department and horticulture departments plays an important role in the implementation of long term conservation plans as botanic gardens can provide the expert advice, practical assistance, horticultural practices, databases and information needed for their conservation and sustainable use.

This edited book is the outcome of an effort to bring in the botanic gardens of India under a network to share their activities and research on conservation and horticulture. A quick scan through the titles of chapters will reveal the intention and purpose of the book. The introduction chapter by the editors gave an overview of the botanic gardens in India and review of a few leading botanic gardens. The book covers important conservation efforts being carried out at AJC Bose Indian Botanic Garden (Howrah), CSIR-Institute of Himalayan Bioresource Technology (Palampur), National Botanic Garden (Trombay), Bhagalpur University (Bhagalpur), Botanic Garden of Indian Republic (Noida), Government Botanical Garden, (Udhagamandalam), Tamil Nadu Agricultural University (Coimbatore), Botanical Survey of India (Northern Circle, Dehradun), Guru Nanak Dev University (Amristar) *etc.* Though the majority of the chapters are on conservation of germplasm and their challenges, there are also chapters on horticulture research being carried out in different botanic gardens, institutes and universities. Hence, the editors are hopeful readers will be able to find articles of their preferred subject area in this book.

The editors would like to extend their sincere thanks to all the contributors for sharing their research and conservation efforts and making this book rich and diverse. The editors are thankful to the Director, CSIR-National Botanical Research Institute, Lucknow for providing facilities and encouragement. Finally, the editors are thankful to the publisher, Astral International Pvt. Ltd. and its staff for their kind co-operation and for publishing this book.

R.K. Roy

J.S. Khuraijam

Contents

Chapter 1

Botanic Gardens of India: Treasure house of plant diversity

J.S. Khuraijam and R.K. Roy

Botanic Garden Division, CSIR-National Botanical Research Institute, Lucknow
E-mail: jskhuraijam@yahoo.com

Abstract

Botanic gardens in India with its varied roles have been instrumental in conserving plant diversity of the country, promotion of environmental education and scientific research in plant sciences and horticulture. Around 150 Botanic Gardens are there in India and most of them are associated with Universities and Institutes. Notable botanic gardens are Acharya Jagadish Chandra Bose Indian Botanic Garden, Howrah; Lalbagh Botanical Garden, Bengaluru; Jawaharlal Nehru Tropical Botanic Garden and Research Institute, Thiruvananthapuram and CSIR-NBRI Botanic Garden, Lucknow. With beautiful landscapes, display of magnificent and colourful plants with labels, flower shows, educational and societal services, these botanic gardens are the source of knowledge for cultivation, utilisation and conservation of plants. Living collections of plants in botanic gardens helps in maintaining a living repository of genetic diversity that can support many activities in conservation and research. Over the years, many economical and ornamental plants have been collected, documented, propagated, and conserved in the botanic gardens. However lack of proper management, expert staffs and funds have led to deterioration of several gardens. Challenges faced by botanic gardens need to be addressed at regional as well as national level and an effective work plan for proper management of Botanic Garden is required.

Keywords: *Botanic garden, Conservation, Plant diversity, India*

INTRODUCTION

The importance of plants for existance of life on earth is immense. Humans depend upon them for the air to breathe, for the food to eat, for shelter, and for medicine. Ninety percent of all terrestrial species on our planet exist in the tropical countries. However, these regions have the highest extinction rate of species and they are disappearing faster than before. Conserving them for your future generations has become responsible duty of all of us. Educating the forest dwellers, urban people and younger generation about the importance and utility of plants to our today-to-day life have become important. It is now necessity to conserve them and use them sustainably.

Botanic Gardens play an important role in conserving plant diversity and protecting them from extinction through plant exploration, propagation, habitat restoration, scientific research, and education. For some threatened species, they are only place for their survival. *Encephalartos woodii* is one such species that had gone extinct in wild and are found only in Botanic Gardens. The function of Botanic Garden is multidimensional and productive. World's foremost research in plant sciences and horticulture are carried out in Botanic Gardens across the globe. These are with a purpose to recreate and understand significance of the aesthetic values of plants in our lives. Moreover, to enrich our knowledge about plant diversity and to conserve them.

Brief History of Botanic Gardens

The history of botanic gardens is linked to the history of botany itself. The world's first botanic gardens were the physic gardens of Italy in the 16th and 17th centuries. The first of these physic gardens was the garden of the University of Pisa which was created by Luca Ghini in 1543. Following this idea of physic gardens, other universities in Europe created such gardens. These gardens of the 16th and 17th centuries were purely medicinal gardens, but the idea and usage of the garden changed over the time to encompass displays of the beautiful, strange, new and sometimes economically important plants being returned from the European colonies and other distant lands. Beginning of exploration and international trade led to the cultivation of new species that were being brought back from expeditions to the tropics. Royal Botanic Gardens, Kew and the Real Jardín Botánico de Madrid played an important role in the promotion of botanical exploration in the tropics and they also helped in establishing new gardens in the tropical regions to cultivate these newly discovered plant species. The British and French established tropical gardens in India and Mauritius for cultivation of commercial crops such as cloves, tea, coffee, breadfruit, cinchona, palm oil as well as chocolate. They introduced rubber plantation to Singapore, teak and tea to India and breadfruit, pepper and star fruit to the Caribbean. In the 18th century, these gardens became more educational in function, demonstrating the latest plant classification systems. Later, with the increased collection of ornamental and diverse groups of plants led to the incorporation of both horticulture and botany in botanic gardens. With the emergence of conservation movement in recent times, botanic gardens played important role in *ex-situ* conservation of threatened species. Their role as refugee for several plant species increased multifold after IUCN encouraged *ex-situ* conservation of threatened plants. IUCN Volunteer members in many botanic gardens are responsible for the success of this movement.

At present, there are around 2000 botanic gardens and arboreta in 150 countries around the world with many more under construction or being planned.

Botanic Gardens in India

In India, there are around 150 botanic gardens devoted to conservation of native plants of the country and exotic species which are of aesthetic and economical values. Some of the botanic gardens are listed in Table 1.1. Most of the Botanic Gardens in India are associated with Universities and Institutes where it serve as an entity of R&D activities besides other services of a botanic garden. A brief overview of four leading Botanic Gardens of India *viz.*, Acharya Jagadish Chandra Bose Indian Botanic Garden, Howrah; Lalbagh Botanical Garden, Bengaluru; Jawaharlal Nehru Tropical Botanic Garden and Research Institute, Thiruvananthapuram and CSIR-NBRI Botanic Garden, Lucknow are given below:

1. Acharya Jagadish Chandra Bose Indian Botanic Garden, Howrah, West Bengal

The Acharya Jagadish Chandra Bose Indian Botanic Garden is one of the old botanic gardens in the world. The garden was founded by Lieutenant-Colonel Robert Kyd of the East India Company in the year 1787. During the reign of East India Company, the garden was called 'Company Bagan'. Later it was renamed 'Royal Botanic Garden, Calcutta' when Queen Victoria took over the East India Company and remained so till India become independent. After impendence, it was renamed as the 'Indian Botanic Garden' which is now known as Acharya Jagadish Chandra Bose Indian Botanic Garden. The Botanic Garden is located on the bank of the river Hoogly and covers an area of about 273 acres. The Garden was developed for the collection of plants, indigenous to the country and for the introduction and acclimatization of plants from foreign parts. It was from these gardens that the tea now grown in Assam and Darjeeling was first developed. The Indian Botanic Garden is most famous for the Great Banyan Tree, having the largest canopy in the world.

Table 1.1: List of Botanic Gardens in India

Sl.No.	*Name of Botanic Gardens*
1	Acharya Jagadish Chandra Bose Indian Botanic Garden, Howrah
2	Agri-Horticultural Society, Madras
3	Aligarh Muslim University Botanic Garden, Aligarh
4	Arboretum, Manali
5	Athreya Research Foundation, Ernakulam
6	Auroville Botanical Gardens, Kottakuppam
7	AVREFA Herbal Garden, Palghat
8	Barapani Experimental Garden, Shillong
9	Bhagalpur University Botanical Garden, Bhagalpur
10	Bhel Garden Complex, Bhel
11	Botanic Garden of Indian Republic, Noida
12	Botanic Garden, Anand
13	Botanic Garden, Bidhan Chandra Krishi Viswavidyalaya, Cooch Behar
14	Botanic Garden, CSIR-Central Drug Research Institute, Lucknow
15	Botanic Garden, CSIR-Institute of Himalayan Bioresource Technology, Palampur
16	Botanic Garden, CSIR-National Botanical Research Institute, Lucknow
17	Botanic Garden, Sirai Taluka
18	Botanical Garden (Motibang), Junagadh
19	Botanical Garden, Agra College, Agra
20	Botanical Garden, Bihar Agricultural College, Bhagalpur
21	Botanical Garden, Delhi University, Delhi
22	Botanical Garden, Dibrugarh University, Dibrugarh
23	Botanical Garden, Dr H.S. Gour Vishwavidyalaya, Sagar
24	Botanical Garden, Guru Nanak Dev University, Amritsar
25	Botanical Garden, Forest Research Institute, Dehra Dun
26	Botanical Garden, Manipur University, Imphal
27	Botanical Garden, Mumbai Institute of Sciences, Mumbai
28	Botanical Garden, Nagaland University, Lumami
29	Botanical Garden, Tamil Nadu Agricultural University, Coimbatore
30	Botanical Garden, NBPGR, Delhi
31	Botanical Garden, S.V. University, Tirupati
32	Botanical Garden, Yogi Vemana University Kadapa
33	Botanic Garden, North Eastern Hill University, Shillong
34	Botanic Garden, Dharwad
35	Brindavan Gardens, Mysore
36	Burdwan University Botanic Garden, Burdwan
37	Calicut University Botanical Garden, Calicut
38	Chessa Botanic Garden, Itanagar

Sl.No.	*Name of Botanic Gardens*
39	Coimbatore Zoological Park Society, Coimbatore
40	Darjeeling Botanic Garden, Darjeeling
41	Department of Botany and Botanic Garden, Bolpur
42	Department of Botany and Botanic Garden, Madras
43	Dept. of Bio-Sciences and Botanic Garden, Jammu Tawi
44	Dept. of Botany and Botanic Garden, Chitrakut
45	Dept. of Botany and Botanic Garden, Guwahati
46	Dept. of Botany and Botanic Garden, Hamdard Nagar
47	Dept. of Botany and Botanic Garden, Jaipur
48	Dept. of Botany and Botanic Garden, Mysore
49	Desert Botanic Garden, Jodhpur
50	Dharma Vana Arboretum, Hyderabad
51	Dhauladhar Botanical Gardens and Arboretum, Palampur
52	Eden Gardens, Calcutta
53	Empress Botanical Gardens, Pune
54	Entomology Research Institute Botanic Garden, Madras
55	Experimental Botanic Garden Andaman Nicobar Circle, Dhanikheri
56	Experimental Botanic Garden Arunachal Field Station, Sankie View
57	Experimental Botanic Garden Arid Zone Circle, Jodhpur
58	Experimental Botanic Garden Central Circle, Allahabad
59	Experimental Botanic Garden Eastern Circle (Shillong), Barapani
60	Experimental Botanic Garden Northern Circle, Pauri
61	Experimental Botanic Garden Northern Circle, Khirsu
62	Experimental Botanic Garden Northern Circle, Dehradun
63	Experimental Botanic Garden Sikkim Himalayan Circle, Gangtok
64	Experimental Botanic Garden Southern Circle, Yercaud
65	Experimental Botanic Garden Western Circle, Mundhwa, Pune
66	Forestry Arboretum, Dhaulakuan
67	Foundation for Revitalisation of Local Health Traditions Botanic Garden, Bangalore
68	Government Gardens, Chaubattia
69	Gujarat University USSC Botanic Garden, Ahmedabad
70	Gurukula Botanical Sanctuary, North Wayanad
71	Herbarium and Botanic Garden, Jammu-tawi
72	Horticultural Experiment and Training Centre, Saharanpur
73	Indian Botanical Garden, Jodhpur
74	Indian Institute of Horticultural Research, Bangalore
75	Indira Gandhi Memorial Botanical Garden, Rae Bareli
76	Institute of Forest Genetics and Tree Breeding, Coimbatore
77	Jawahar Kunj Garden, Barrackpore
78	Jawaharlal Nehru Botanic Garden, Rumtek

Sl.No.	*Name of Botanic Gardens*
79	Jawaharlal Nehru Tropical Botanic Garden and Research Institute, Thiruvananthapuram
80	Jiwaji University Charak Garden, Gwalior
81	Jubilee Park, Jamshedpur
82	Karnatak University Botanical Garden, Dharwad
83	Kodaikanal Botanic Garden, Palayamkottai
84	Kotla Vijayabhaskara Reddy Botanical Garden, Hyderabad
85	Kulbhaskar Ashram Post Graduate Botanic Garden, Allahabad
86	Lalbagh Botanical Garden, Bangalore
87	Lloyd Botanic Garden, Darjeeling
88	Madurai Kamraj University Botanic Garden, Madurai
89	Magadh University Botanical Garden, Bodh Gaya
90	Maharashtra Nature Park Society, Maharashtra
91	Malabar Botanical Garden and Institute for Plant Sciences, Kozhikode
92	Marathwada University Botanic Garden, Aurangabad
93	Meerut University Botanical Garden, Meerut
94	Narendra Narayan Park, Cooch Behar
95	National Botanic Garden, Bhaba Atomic Research Centre, Mumbai
96	National Orchidarium and Botanic Garden, Shillong
97	NHCS - Nelliyampathi Hills Conservation Society, Padagiri
98	Orchid Preservation Centre Khonghampat,Imphal
99	Osmania University Botanic Garden, Hyderabad
100	Padmaja Naidu Himalayan Zoological Park, Darjeeling
101	Pampavana Garden, Munirabad
102	Panjab University Botanical Garden, Chandigarh
103	Peermade Development Society, Azhutha
104	Punjabi University Botanic Garden, Patiala
105	Rajiv Gandhi University Botanical Garden, Rono Hills
106	Ravi Sankar University Botanical Garden, Raipur
107	Regional Plant Resources Centre, Bhubaneswar
108	Rhododendron Arboretum, Gangtok
109	Roxburgh Botanic Garden, Allahabad
110	Sahyadri Botanical Garden, Kerala
111	Sarabhai Foundation Botanic Gardens, Ahmedabad
112	Satpuda Botanic Garden - College of Agriculture, Nagpur
113	Saurashtra University Experimental Garden, Rajkot
114	Sim's Park Coonoor, Coonoor
115	Sri Chamarajendra Park, Bangalore
116	State Botanical Garden, Barang
117	State Horticultural Farm, Kallar

Sl.No.	Name of Botanic Gardens
118	State Horticultural Research Station, Krishnanagar
119	Surya Kunj, Almora
120	Tezpur University Botanical Garden, Tezpur
121	The Agri-Horticultural Society of India, Calcutta
122	The Botanical Garden, Gandhi Krishi Vignana Kendra, Bangalore
123	The Bryant Park, Kodaikanal
124	The Pondicherry Botanical Garden, Pondicherry
125	Tropical Forest Research Institute, (ICFRE), Jabalpur
126	University Botanic Garden, Coimbatore
127	University of Poona Botanic Garden, Pune
128	Utkal University Botanic Garden, Bhubaneswar
129	Uttarayan Complex and Gardens, Shantiniketan
130	Veermata Jijabai Bhosle Udyan and Pranisangrahalaya, Mumbai

Germplasm Collections

Acharya Jagadish Chandra Bose Indian Botanic Garden has one of the finest collections of native and exotic species. There are around space 14,000 plants under 1,300 species. The garden has largest collection of palms representing about 116 species under 53 genera and 8 subfamilies of Arecaceae (Palmae). This is one of the largest refugia of palms in South Asia. Besides that there are good collections of Bambuseae (26 spp.), Orchidaceae (80 spp.), Jasmines (25 spp.), water lilies (30 cvs of 4 spp.), Pandanus, Bougainvillea (148 cvs.), Citrus, succulents (100 spp.) and medicinal plants (450 spp.). Many of the species are economical having great potentiality in agriculture, industries as well as in domestic use and utility. Special emphasis is centered on the conservation and multiplication of rare, endangered and endemic species under *ex-situ* condition in the Garden.

2. Lalbagh Botanical Garden, Bengaluru

The Lalbagh Botanical Garden, Bangalore is of royal origin and was started initially as a private garden in an area of 40 acres by Hyder Ali, one of the most famous rulers of old Mysore in 1760. Initially designed in Mughal style, on the model of an extensive garden at Sira in Tumkur near Bangalore, this garden was further developed by Hyder Ali's son Tipu Sultan and subsequently by the British and Indian doyens of horticulture by extension of area and addition of a number of plant species. Lalbagh was given the status of a Government Botanical Garden in 1856, and since then, it has been an internationally renowned centre for scientific study of plants and botanical artwork and also conservation of plants. Formal and informal styles dominate the garden in perfect harmony, which is a testimony to the beauty of nature. Today, the garden is a lush green paradise with an area of 240 acres in the heart of the city.

Out of the many artistic structures in Lalbagh, the Glass House is the most famous. It was built in 1889 during the administration of Sri John Cameron to

Glass House, Lalbagh Botanical Garden.

commemorate the visit of Prince of Wales. Designed on the lines of the Crystal Palace of England, it was intended for acclimatizing the exotic plant specimens. Today, as the jewel of Lalbagh, it is the centre stage for holding the famous biannual flower shows.

Germplasm Collections

The botanical garden is enriched with numerous native and exotic flora of wide ranging diversity, use and interest. More than 2000 species of plants are found in Lalbagh. Some of the exotic species in the Garden are *Amherstia nobilis, Averrhoa bilimbi, Bixa orellana, Brownea grandiceps, Castanospermum australe, Cola acuminata, Corypha umbraculifera, Couroupita guianensis, Cycas revoluta, Eriobotrya japonica, Swietenia mahagoni etc.* Indigeneous species such *as Artocarpus heterophyllus, Bombax ceiba, Butea monosperma, Cassia fistula, Cycas circinalis, Dillenia indica, Lagerstromia speciosa, Magnolia champaca, Mesua ferrea etc.*, can be seen. In addition, a number of ornamental and economic plant species both of exotic and indigenous origin can be found in Lalbagh.

3. Jawaharlal Nehru Tropical Botanic Garden and Research Institute, Thiruvananthapuram, Kerala

Jawaharlal Nehru Tropical Botanic Garden and Research Institute (JNTBGRI) was found in 1979 with the objective of establishing a Conservatory Botanic Garden of tropical plant resources in general and of the country and the Kerala state in particular. It also undertakes research programmes for the sustainable utilization of the resources. It is the only organization in India, which maintains a 300 acre conservatory garden for the wild tropical plant genetic resources of the country, besides a well integrated multidisciplinary R&D system dealing with conservation, management and sustainable utilization of tropical plant resources. During the past 30 years, it has flourished into one of the premier R&D organization in Asia, devoted to conservation and sustainable utilization of tropical plant diversity. The institute is recognized as a 'National Centre of Excellence in *ex-situ* conservation and sustainable utilization of tropical plants diversity' by the Minister of Environment and Forests, Government of India and the Centre of Science and Technology of Non-Aligned and other Developing Countries (NAM S and T Centre).

Bambusetum at JNTBGRI

Germplasm Collections

Over 3500 trees belonging to 700 species are conserved in the Arboretum of JNTBGRI. Though majority of the plants are from Western Ghats region, about 20 per cent are exotic plants. Some of the notable species are *Antiaris toxicaria, Pterygota alata, Baccaurea courtallense, Adansonia digitata, Amherstia nobilis, Gustavia augusta*. The fern collection is one among the largest collection of living ferns within the country consisting of 150 species in 70 genera belonging to 30 families. In the Palmetum over 150 palm species from different parts of the globe are grown. JNTBGRI have a collection of about 350 species of succulents, more than 40 species/cultivars of bromeliads, 45 species of jasmines, 16 species of cycads and 24 species of conifers ***etc.***

4. Botanic Garden, CSIR-National Botanical Research Institute, Lucknow

CSIR-NBRI Botanic Garden is one of the oldest and historical Botanic Gardens in India. The Garden is known for its immense contributions towards conservation and sustainable utilization of important plant resources of economic, ornamental, horticultural, biological, ecological, educational and recreational values. The Botanic Garden is actually the historical 'Sikander Bagh' laid out around 1800AD as a royal garden by the 'nawabs' (King) of Lucknow. It is located in the heart of Lucknow, the capital city of Uttar Pradesh state. Spread over an area of 65 acres with full of greenery and vast collection of plant diversity, the garden serves as lung of the city. This Botanic Garden serves as a National Facility with four main functions *viz.* conservation, education, scientific research and display of plant diversity in plant houses and arboreta. It is designed to conserve the indigenous and exotic flora and fulfils the basic function of making available for study, research and use, at one place, a wide diversity of trees, shrubs, climbers and other plant species. The Botanic Garden is a living repository of over 5000 taxa/cultivars of various groups of native and exotic plants. The garden has an excellent collection of ornamental crops, trees, houseplants, medicinal plants, cycads, palms, ferns, bryophytes, bonsai, water lilies, cacti and succulents. There are several thematic gardens and plant houses where plants are displayed for educational and aesthetic purpose. Considering the rich plant diversity in the Botanic Garden, the Institute has been designated as Living National Repository by the National Biodiversity Authority, Chennai. Besides, this Botanic Garden has also been recognized as a Lead Botanic Garden by the Ministry of Environment, Forests and Climate Change, Government of India for enhancing the *ex-situ* conservation activities of threatened species.

Germplasm Collections

Botanic Garden has an excellent germplasm collection of 202 cultivars of Bougainvilleas; 80 cultivars of Gladiolus; 250 cultivars of Chrysanthemums; 50 cultivars of Canna, 125 cultivars of Roses; a Fern House with 65 species of ferns and fern allies; an arch shaped Conservatory for tropical and subtropical plants with 300 species/cultivars; a Cactus and Succulent House with 200 species/varieties; a Palm House with 52 species; a Cycad House (70 species), an Orchidarium with 120 species and a Moss House (20 species) The Arboretum has around 400 species of trees and shrubs.

Arboretum

Nearly 400 species of trees are conserved in the aboretum. These are planted in systematic order and properly labelled. Some notable indigenous and exotic tree species are: *Adansonia digitata, Alstonia scholaris, A. macrophylla, Annona muricata, Boswellia serrata, Bischofia javanica, Butea monosperma, Chorisia insignis, Cinnamomoum camphora, Coccoloba uvifera, Dillenia indica, Diospyros malabarica, Dalbergia lanceolata, Ficus benghalensis, F. krishnae, Litsea glutinosa, Oroxylum indicum, Pterocarpus marsupium, Saraca declinata, Shorea robusta, Santalum album, Strychnos nux-vomica, S. potatorum, Syzygium jambos, Tabebuia palmeri, Tecomella undulata* and *Tectona grandis*.

Conservatory

A horseshoe-shaped plant house, with an area of 1,370 sq.m, serves as a conservatory for house plants and species from tropical and sub-tropical climates. There are nearly 300 species/cultivars of ornamental, wild and cultivated plants in the conservatory. Some interesting plants are: *Bambusa ventricosa, Costus pictus, Encephalartos villosus, Microcycas calocoma, Hoya carnosa, Fatsia japonica, Ginkgo biloba etc*. There is also a large collection of *Aglaonema, Alocasia, Anthurium, Asparagus, Calathea, Chlorophytum, Codiaeum, Dieffenbachia, Dracaena, Peperomia, Philodendron, Pandanus* and *Syngonium*.

Cactus and Succulent House

A pagoda-shaped glasshouse, with an area of 284 sq.m, holds the germplasm collection of about 250 species or varieties of cacti and succulent plants from arid regions. Some of the important genera under ex-situ conservation in this glass house are - *Adenium, Agave, Aloe, Astrophytum, Cereus, Cissus, Cleistocactus, Coryphantha, Dasylirion, Disocactus, Echinocactus, Echinocereus, Echinopsis, Epiphyllum, Euphorbia, Ferocactus, Gasteria, Graptopetalum, Haworthia, Huernia, Hylocereus, Kalanchoe, Lithops, Mammillaria, Opuntia, Pereskia, Sansevieria, Sedum, Selenicereus, Stapelia, Stenocereus* and *Yucca*. The plants are aesthetically laid out in informal beds and landscaped with pebbles and rocks in a fascinating way. The plant house also conserves rare and threatened species such as *Aloe harlana, Coryphantha maiz-tablasensis, Dracaena draco, Echinocactus grusonii* and *Euphorbia cylindrifolia*. Around 25 species in the plant house are in the IUCN Redlist. The major attraction of the Cacti and Succulents House is the majestic living specimen of *Welwitschia mirabilis*. This Botanic Garden is the only custodian of this bizarre plant in entire South Asia. Important Indian succulents have also conserved in the plant house. Medicinally important succulents such as *Euphorbia neriifolia, Plectranthus amboinicus* and *Aloe vera* have been displayed and conserved.

Palm House

The Palm House covers 765 sq.m and contains the Arecaceae germplasm collection. 52 species of palms are displayed in pots of various sizes and in the ground. Some interesting species are: *Arenga pinnata, Areca catechu, Cocos nucifera, Caryota urens, C. mitis, Chamaedorea elegans, C. stolonifera, Chrysalidocarpus lutescens, Daemonorops kunstleri, Elaeis guineensis, Licuala grandis, L. spinosa, Livistona chinensis,*

L. cochin-chinensis, Mascarena verschaffeltii, Phoenix reclinata, Ptychosperma macarthuri, Thrinax barbadensis and *T. excelsa*.

Fern House

This pyramidal house, with an area of 400 sq.m, holds the germplasm collection of more than 100 fern species from India and abroad, including: *Adiantum capillus-veneris, A. hispidulum, Blechnum occidentale, Bolbitis heteroclita, Diplazium esculentum, Drynaria quercifolia, Equisetum debile, Lygodium flexuosum, Microsorium alternifolium, Nephrolepis cordifolia, N. duffii, N. tuberosa, Ophioglossum reticulatum, Psilotum nudum, Pteris cretica 'albolineata'* and *P. vittata*.

Orchidarium

120 species of orchids are conserved in this plant house. Some notable rare species are - *Paphiopedilum spiceranum, Paphiopedilum insigne, Vanda coerulea, Renanthera imschootiana, Dendrobium aphyllum, Vanda tessellata* and *Eulophia nicobarica*. The orchids were collected from natural habitats of six states *viz*., Assam, Bihar, Jharkhand, Meghalaya, Manipur, Nagaland, Odisha and Uttarakand. Some of the species conserved in the plant house are - *Acampe praemorsa, Arundina graminifolia, Bulbophyllum crassipes, Coelogyne cristata, Cymbidium bicolor, Dendrobium aphyllum, Dendrobium herbaceum, Dendrobium kentrophyllum, Dendrobium moschatum, Dendrobium polyanthum, Gastrochilus inconspicuous, Pelatantheria insectifera, Rhynchostylis retusa etc*.

Pioneering Effort: *Ex-situ* Conservation of Cycads in India

Cycads are relict group of plants which dominated the plant kingdom during Jurassic Period. These ancient plants are the most threatened group of organisms on earth with almost 64 per cent of cycads are at the risk of extinction. Four species of cycads are now extinct in wild and can only be seen in botanic gardens. They are threatened in wild due to clearing of forest, unsustainable harvesting of cycad seeds and forest fire. Cycads have several cultural and economical importance, and the people living near the natural populations used cycads for different purposes *viz*., food, socio-cultural activities, decoration and horticultural uses.

In India, there are 13 species of *Cycas* that grow naturally in the Western Ghats,

Cycad House, CSIR-NBRI, Lucknow.

Eastern Ghats, North East India, and Andaman and Nicobar Islands. CSIR-National Botanical Research Institute (NBRI), Lucknow has taken up the task of conserving this rare group of plants. The Institute has the largest collection of cycads (70 species) in South Asia which include 9 species of Indian *Cycas viz. Cycas annaikalensis, C. circinalis, C. beddomei, C. nayagarhensis, C. orixensis, C. pectinata, C. sphaerica, C. swamyi* and *C. zeylanica.* The Institute has successfully developed seed germination techniques of six species of Indian *Cycas* and efforts are being made to develop for remaining species.

Besides Indian species, nineteen species of cycads have been successfully propagated in the Cycad House through seed germination *viz. Cycas revoluta, C. micronesica, C. nayagarhensis, C. nongnoochiae, C. orixensis, C. pachypoda, C. pectinata, C. riuminiana, C. schumanniana, C. seemannii, C. silvestris, C. tansachana, C. wadei, C. zeylanica, Macrozamia moorei, Zamia erosa, Z. furfuracea, Z. loddigesii* and *Z. pumila.* Seedlings of *Zamia loddigesii, Z. pumila* and *Z. furfuracea* were raised through germination of viable seeds developed by artificial pollination. Artificial pollination of *Microcycas calocoma, Dioon spinulosum* and *Cycas revoluta* is being carried out. Besides these, seven species were propagated through vegetative propagation.

Conclusion

Lack of framework for proper administration and expertise in plant conservation has been major challenges for Botanic Gardens of the country. Over-collection of plants in Botanic Gardens just for the sack of enrichment of germplasm without proper maintenance and vision may result in depletion of wild populations. Staff of many Botanic Gardens are not well versed with the concept of documentation of living collections. Skills of both scientifically trained plant conservationist and the traditionally trained horticulturist need to be developed and recognised. Not only having enough number of well-trained staff, lack of funds is also another major bottle neck in management of botanic gardens and carrying forward their conservation efforts. All these challenges need to be addressed at regional as well as national level and a concrete framework for proper management of Botanic Garden is required. Ministry of Environment and Forests, Government of India initiated a scheme in 1992 entitled "Assistance to Botanic Garden Scheme" for conservation of rare endemic plants of the country. The botanic gardens under this scheme are periodically monitored with the help of Botanical Survey of India (BSI). This scheme is a positive step in supporting the botanic gardens to speed up the plant conservation efforts and improve their over-all facilities.

Botanic Gardens of India are in urgent need for a collaborative network. It's high time for the policy makers and Botanic Gardens to work together to bring in all the Botanic Gardens of the country under one umbrella and to accelerate measures for conservation as well as management.

Acknowledgements

The authors are thankful to the Director, CSIR-NBRI for the facilities and encouragement. The authors are also thankful to staff of the Botanic Garden for their inputs and co-operation.

Further Readings

BGCI. 2019. Botanic Gardens Conservation International. (https://www.bgci.org/)

Cameron, J. 1891. Catalogue of plants in the botanical garden, Bangalore and its vicinity. Mysore Govt. Central Press.

CSIR-NBRI Botanic Garden. (https://nbri.res.in/bgarden.php)

Maunder, M. 1994. Botanic gardens: future challenges and responsibilities. Biodiversity and Conservation 3: 97-103.

MoEF&CC. 2017. Financial Assistance under "Assistance to Botanic Garden Scheme" (http://www.moef.gov.in/call-for-proposals-for-financial-assistance-under-assistance-to-botanic-garden-scheme/)

Mohanan, N. 2009. Thirty years of TBJRI. In Garden and People 2009, National Seminar on Botanic Gardens at Thiruvananthapuram, Kerala. Souvenir, p13-15

Mondial, M.S., S. Gantait, S. Nandi, T. Chakrabortty, Krishna Das & S. Bandyopadhyay. 2009. AJC Bose Indian Botanic Garden, Howrah - A guide to visit garden and herbarium. ENVIS Newsletter 14(1): 5-7.

Pandey, D.S. & Chowdhery, H.J. 2007. Plants of Indian Botanic Garden. Bishen Singh Mahendra Pal Singh, Dehradun.

Singh, P. & Dash, S.S. 2017. Indian Botanic Gardens: Role in Conservation. Botanical Survey of India, Kolkata.

Smith, P. 2019. The challenge for botanic garden science. Plants, People, Planet. 1: 38-43.

TBGRI. 2005. Tropical Botanic Garden and Research Institute. Science India 35-38. (https://jntbgri.res.in/files/press/scienceindia_july2005.pdf)

Maunder, M. 1994. Botanic gardens: Future challenges and responsibilities. Biodiversity and Conservation. 3(2): 97-103.

Chapter 2

Botanical Garden of CSIR-Institute of Himalayan Bioresource Technology, Palampur (H.P.): An Introduction

Brij Lal*, Gopichand and Alka Kumari

Department of High Altitude Biology, CSIR-Institute of Himalayan Bioresource Technology, Palampur – 176 061, H.P.
**E-mail: brijlal@ihbt.res.in*

Abstract

CSIR-Institute of Himalayan Bioresource Technology (IHBT) Palampur is involved in exploration, mapping, domestication, multiplication and conservation of economic, rare, endangered and threatened (RET) plants of western Himalayan region. In order to conserve and multiply different valuable plant species of western Himalayan region through botanical garden, Scientists at CSIR-IHBT Palampur took initiatives to establish a botanical garden and started exploration and collection of plants from different locations of Himalayan region. As a result of extensive explorations and plant colelction, around 400 plant species of trees, shrubs, herbs, climbers, bamboos, gymnosperms and pteridophytes (ferns and fern allies) were introduced in the existing botanical garden of the Institute. Among the interesting species introduced in the botanical garden are Aesculus, Cedrus, Crataegus, Ginkgo, Platanus, Taxodium and Taxus (tree species); Bauhinia, Clerodendrum, Cryptolepis, Thunbergia, Tinospora (climbers); Berberis, Incarvillea, Jasminum, Valeriana, Zanthoxylum (herbs/shrubs); and Adiantum, Diplazium, Polystichum, Pteridium, Pteris species (pteridophytes). The details about the current status of plant diversity maintained and future strategies for conservation and sustainable utilization of plant resources through botanical garden have been discussed.

Keywords: *Botanical garden, Western Himalaya, Conservation, RET taxa.*

INTRODUCTION

Conservation of threatened and economic plant resources has become a matter of great concern in the world today. Amongst the resources gifted by nature, the flora of Himalayan region provides an everlasting and interesting field of investigation. Plant diversity and uniqueness of the flora in varied ecosystems have retained sound, clean and aesthetic environment of the region. Hence, the flora has remained always a matter of great curiosity to the botanists, tourists and naturalists. Today, several plant species growing in their natural habitats are facing varied degree of threats and some of them have become rare due to continuous extraction from nature by different groups users for manufacturing different products. Moreover, many people from the urban areas generally remain away from closer contact with the fascinating flora of Himalaya. To introduce such group of people with the natural flora of Himalaya without going remote distant areas and to make them aware about the economic importance of Himalayan plant resources and their conservation, establishment of a botanical garden at Palampur conditions has been deeply realized. Therefore, keeping in view the above back ground, the Scientists at CSIR-Institute of Himalayan Bioresource Technology (IHBT) Palampur took initiatives to develop a botanical garden at Palampur aimed with the following objectives:

1. Improvement and strengthening of existing botanical garden.
2. *Ex-situ* conservation and multiplication of threatened and economic plant species.
3. Reintroduction of selected plants in nature in collaboration with forest department.
4. Promoting collaborative research programmes.
5. Organizing education and public awareness programmes.
6. Promoting ecotourism in the area.
7. Developing linkages with the network of botanical gardens.

Facilities

CSIR-Institute of Himalayan Bioresource Technology (IHBT) Palampur, a constituent laboratory of CSIR India, has established earlier as CSIR at Palampur. The laboratory started functioning in the year 1984 with a mission to provide an industrial R&D base for the establishment, upgradation and sustainable management of bioresources in the Himalayan region through exploration, agrotechnology, processing technology and biotechnology. The laboratory is located between 32°06′00″ - 32°07′48″ N latitude and 76°31′48″ - 76°34′12″E longitude at the picturesque town of Palampur, Kangra district, H.P. Perched in the lap of majestic snow clad mountain of Dhauladhar, the institute is spread over 222 acres area (Figure 2.1A). The institute is dedicated in providing R&D services on economic bioresources in western Himalayan region leading to value added plants, products and processes for industrial, societal and environmental benefits. The institute is also involved in survey, mapping, collection, conservation, propagation, domestication and documentation of plant resources of western Himalayan region. As a result of intensive efforts put by the scientists of the Institute, a number of world class R&D facilities like database on plants of western Himalaya, chemistry, tissue culture and molecular biology labs, tea processing unit, NMR and viral testing facilities were created. In addition, infrastructure facilities like internationally recognized herbarium (acronym PLP), bamboo museum, tea, fernery, rose and botanical gardens and green houses supporting institute's R&D activities were created over the period of time. Thus, the institute has technical capabilities to domesticate, propagate and conserve threatened and economically important species of the region.

Botanical Garden at a Glance

The history of conserving plant genetic resources by human beings is as old as civilization. Collection, multiplication and domestication of seeds, tubers/ rhizomes and other planning materials through botanical gardens is one of the best strategies of plant conservation (*https://www.bgci.org/resources/bgs_in_conservation*). The CSIR-IHBT Palampur is involved in exploration, collection, domestication, multiplication and conservation of economic, rare and threatened plants of western Himalayan region for last the two decades. In order to conserve and multiply various economic and threatened plant species of Himalayan region through botanical garden, initiatives were taken to establish a botanical garden (Figure 2.1B) and started exploration and collection of plants from different locations of Himalayan region. Spread over an area of 20 acres, the existing botanical garden is located at a transit zone (1300m amsl) which provides a congenial environment to grow the plants found in both subtropical and temperate regions. A glimpse of the same is given in Figure 2.1 (Figure 2.1E-2.1H). The garden has been designed to develop as a potential and an accessible repository of live plant resources found in remote and high altitude areas which are beyond the reach of a common man. Thus, it would form strong base for plant conservation and environmental protection. The existing botanic garden is comprised of different sections like arboretum, bamboosetum, fernery, rose and herbal garden, herbarium, and bamboo museum. At present, the garden is represented by around 400 plant species (127) ferns and fern allies, 120 herbs/shrubs, 110 trees and climbers and 28 bamboo species) maintained in different sections of the garden as given below:

Arboretum

The arboretum harbors around 110 trees, shrubs and woody climbers collected from different parts of the country. Among the species maintained in the arboretum are *Agathis robusta, Albizia* spp., *Alnus nitida, Aesculus indica, Bauhinia vahlii, B. variegate, Cedrus deodara, Crataegus oxyacantha, Erythrina blackei, Ginkog biloba, Mallotus philippensis, Myrica esculenta, Pinus roxburghii, Platanus orientalis, Quercus leuchotrichophora, Sapidnus mukorosi, Sapium* spp. *Taxodium distichum, Taxus baccata, Toona ciliata etc.*

Fernery

In recent past R&D work on pteridophytes were initiated at IHBT Palampur (Ahuja and Singh 2012). Consequently, a conservatory (fernery) has also been established which provides shelter to about 127 ferns and fern allies (pteridophytes)

Figure 2.1: A Glimpse of IHBT Botanical Garden.

A. Field view, B. A view of Botanical garden, C. *Ginkgo biloba* plantation, D. Ferny, E. *Erythrina blackei*, F. *Prinsepia utilis*, G. *Bauhinia variegata*, H. *Reinwardtia indica*.

collected from different locations (Figure 2.1D). The collection is represented by some edible, rare, medicinal and ornamental fens like *Adiantum* spp., *Coniogramme*, and *Cyrtomium* spp. *Cyathea spinulosa, Diplazium esculentum, D. maximum, Lygodium flexuosum, Nephrolepis cordifolia, Osmunda claytoniana, Polypodium* spp., *Polystichum squarrosum, Pteridium aquilinum, Pteris cretica, P. vittata etc.*

Herbal Garden

A collection of around 100 species of medicinal and aromatic plants maintained in the herbal garden includes some rare and threatened plants collected from western Himalayan region. Some of the herbs maintained in the herbal garden are *Acorus calamus, Adhatoda zeylanica, Aloe vera, Berberis lycium, Cryptolepis buchnanii, Curcuma aromatic, Incarvillea emodi, Prinsepia utilis, Tinospora cordifolia, Valeriana jatamansi, Vitex negundo, Viola odorata, V. pilosa, Withania somnifera, Zanthoxylum armatum etc.*

Herbarium

There is a close relation between a garden and herbarium. At IHBT, there is an internationally recognized herbarium (acronym PLP). This is the only recognized herbarium in the state of Himachal Pradesh. It houses around 22000 specimens belonging to 2000 species mainly representing 3 groups of plants *i.e.* Angiosperms, Gymnosperms, and Pteridophytes. The collection lying with this herbarium mostly represents flora of western Himalayan region.

Bambusetum

For raising bamboos in nurseries and their distribution to different localities all across the country, the Institute has established a good collection of bamboo germplasm. There are 28 species of bamboos in the collection including edible and ornamental bamboos collected from different parts of the country and abroad also. A few of them are *Bambusa balcoa, B. bambos, B. nigra, B. nutans, B. pallida, B. tulda, Dendrocalamus asper, D. giganteus, D. hamiltonii, D. strictus, etc.* Among the species like *B. bambos, Dendrocalamus asper, D. hamiltonii* were multiplied and supplied to different states in the country for plantation and generating materials for making artifacts, basketry, and fencing purpose.

Bamboo-Museum

The Institute has a unique bamboo museum which also called as 'Bamboo House'. This is the first bamboo museum in the country (Thapliyal, 2011). The building material used for its construction is only bamboo material. It displays useful artifacts and products made of bamboos collected from different places for display purpose. In addition, a number of articles and R&D products developed at IHBT are also kept for display in the museum for the visitors.

Besides above developments, efforts were made by various workers at IHBT to conserve a number of rare, threatened and ornamental plants facing a varied degree of threats. Among threatened plants, *Jasminum parkeri* Dunn an endemic and endangered species rediscovered almost after hundred years (Lal *et al.*, 2014), has been introduced in the botanical garden of the institute. In addition, introduction and conservation of wild roses and threatened ornamental plant known as *Incarvillea emodi* were carried out by the scientists at IHBT (Dhyani *et al.*, 2010, 2014). The scientists at IHBT also succeeded in multiplying *Ginko biloba,* a valuable medicinal plants known as living fossil. Today, IHBT is the only institution in the country where a large number of plants (20000 plants) of *G. biloba* are maintained (Figure 2.1C). As a result of extensive plant survey, several rare and threatened plants of Himalayan region were collected and being conserved in the existing botanical garden of the institute. Among the rare and threatened plants conserved at IHBT are given in the Table 2.1.

Table 2.1: Some Rare and Threatened Plants Introduced and Conserved in IHBT Botanical Garden, Palampur

Name of Species	*Family*	*Conservation Status*
Aconuitum heterophyllumn Wall. ex Royle	Ranunculaceae	CR
Dioscorea deltoidea Wall. ex Griseb.	Dioscoreaceae	EN
Cyathea spinulosa Wall. ex Hook.	Cyatheaceae	EN
Ginkgo biloba Linn.	Ginkgoaceae	EN
Hedychium spicatum Ham. ex Smith	Zingiberaceae	VU
Incarvillea emodi Chatterjee	Bignoniaceae	CR
Jasminum parkeri Dunn	Oleaceae	CR
Picrorhiza kurrooa Royle ex Benth.	Scrophulariaceae	EN
Podophyllum hexandrm Royle	Podophyllaceae	CR
Taxus wallichiana Zucc.	Taxaceae	EN
Valeriana jatamansi Jones	Valerianaceae	VU

It is worth to mention that the present activities not only remained restricted to plant collection and their introduction in the botanical garden, but efforts were made to develop cultural and cultivation practices for some valuable medicinal and aromatic plants. For examples cultivation practices have been developed for *Curcuma aromatic* (Gopichand *et al.*, 2006, 2009), *Valeriana jatamansi* (Singh *et al.*, 2000), *Ginkgo biloba* (Prakash *et al.*, 2002) *etc.*

Future Strategies

1. Strengthening infrastructure facilities.
2. Promoting interdisciplinary approach to conserve and multiply highly threatened and economically important plant species.
3. Development and introduction of new sections for special plant groups like Palms, Zingibers, Orchids, Cactus and other interesting groups.
4. Promoting collaborative research.
5. Back to nature and linkages with forest department.
6. Organizing programmes for public awareness.
7. Promoting ecotourism for sustaining garden activities.
8. Developing linkages with botanical gardens network.

Conclusion

Conservation of rare, threatened, economically important and ornamental plants though botanical garden of any region will not only save these plant species from extinction but it also provides opportunity to understand the nature and behavior of plant in *ex-situ* conditions. Subsequently, people living in urban areas get acquainted with the plant resources of remote and high altitude areas through the visit of botanical gardens. Moreover, botanical gardens are the best ways of creating awareness about the conservation and sustainable utilization of plant resources. In addition, botanical garden could be a best source of income generation through ecotourism and value addition to certain promising plants or by selling plant materials to the visitors.

Acknoweldgements

The authors are thankful to the Director, CSIR-IHBT Palampur for the facilities. The funding agencies like CSIR, DBT, DST, MoEF and CC Govt. of India New Delhi, are duly acknowledged for the financial support under different sponsored R&D programmes. Thanks are due to the people who were/are directly or indirectly involved in the development of botanical garden at IHBT.

References

Ahuja P.S. and Singh R.D. 2012. R&D on pteridophytes at CSIR-IHBT Palampur (H.P.). *Indian Fern Journal* 29:269-271.

Dhyani D. and Singh S. 2010. Domestication and conservation of *Incarvillea emodi*: a potential ornamental wild plant. *Indian Journal of Agricultural Sciences* 80(2): 182-185.

Dhyani D. and Singh S. 2014. Potential wild rose germplasm of western Himalayas-Conservation, evaluation and registration. *Indian Journal of Agricultural Sciences* 84(2): 229-235.

Gopichand, Singh R.D., Kumar A, Meena R.L. and Ahuja P.S. 2009. Current status of *Ginkgo biloba* L. in India. *The Indian Forester* 135(11): 1588-1593.

Gopichand, Singh R.D., Meena R.L., Singh MK, Kaul V.K., Lal B., Acharya R., and Prasad R. 2006. Effect of manure and plant spacing on crop growth, yield and oil-quality of *Curcuma aromatica* Salisb. in mid hill of western Himalaya. *Industrial Crops and Products* 24:105–112.

Lal B, Datta A, Parkash O and Singh RD 2014. Rediscovery of *Jasminum parkeri* Dunn: an endemic and endangered taxon from the western Himalaya, India. *Biodiversity Research and Conservation* 34: 11.

Prakash O., Nagar P.K., Lal B. and Ahuja P.S. 2002. Effect of auxins and phenolics acids on adventitious rooting in semi-hard wood cuttings of *Ginkgo biloba. Journal of Non-Timber Forest Products* 9(1/2): 47-49.

Singh R.D., Ahuja P.S., Nagar P.K., Kaul V.K., Singh B, Lal B., Vats S.K., Yadav, P. and Mishra S. 2000. Effect of manuring and shade on yield and quality of *Valeriana wallichi. Journal of Medicinal and Aromatic Plant Sciences* 22(1): 669-670.

Thapliyal, J. 2011. India's first bamboo museum to open doors, The Tribune, 26 September 2011, Tribune News Service, Dehradun.

Chapter 3

Primary Hemi-epiphytes: A Challenge for *Ex-situ* Conservation in AJC Bose Indian Botanic Garden, Howrah – A Case Study

Basant Kumar Singh, Arabinda Pramanik and C.M. Sabapathy*

AJC Bose Indian Botanic Garden, Botanical Survey of India, Howrah – 711 103
**E-mail: singhbk2007@yahoo.co.in*

Abstract

Acharya Jagadish Chandra Bose Indian Botanic Garden (AJCBIBG), Howrah erstwhile known as Company Bagan or Royal Botanic Garden or Indian Botanic Garden was established by Colonel Robert Kyd in 1787. Spanning over an area of 273 acres, this garden has served as the introductory platform for many economically important plants that share the major part of commerce and trade in present India. Actively involved in ex-situ conservation for last 229 years, AJCBIBG is a repository of 14,122 plants under 1,377 species mostly of threatened and endemic nature. But in due course of time, it is observed that some of these plants are getting replaced by different kinds of Primary Hemi-epiphyte species. Detailed field study for last two consecutive years shows that the matter has really taken a serious shape of great concern. Eventually each and every introduced plant in a botanic garden designated exclusively as ex-situ conservation site, is important from either point of view. But the way these primary hemi-epiphytes are establishing themselves on some of the introduced trees in the garden and gradually nibbling the plants is a matter of great concern for conservation and very existence of the host plants (Phorophytes).

Present work dealt with 163 such cases in AJCBIBG ranging from the new intruders to well establish primary hemi-epiphytic Ficus trees belonging to family Moraceae and highlight the way the host plants are strangulated and finally swallowed up in due course of time. Some remedial course of action has been suggested to manage the menace.

Keywords: *Ex-situ conservation, Primary Hemi-epiphyte, Phorophyte.*

INTRODUCTION

Acharya Jagadish Chandra Bose Indian Botanic Garden (AJCBIBG) is one of the biggest and oldest botanic garden in South-East Asia spread over 273 acres of land harbours around 1377 plant species including rare, threatened, endemic and economically important plants (Debnath *et al.*, 2014). In its early days since inception in 1787 the garden had acted as the introductory platform for many economic plants in India, mention may be made of Tea, Cinchona, Rubber, Mahogany, *etc.* Well before the Rio Earth Summit *i.e.* way back in 1987 AJC Bose Indian Botanic Garden, Howrah erstwhile Company Bagan had adopted a well planned decision in its 'Annual Action Plan' for introducing Rare, Endangered and Threatened (RET) indigenous plant species, prior to their extinction from the wild and to multiply and conserved them so as to check the loss of species. This is evident from large scale introduction of plants in AJCBIBG from different corner of the country since long back. During the period of 1987 to 2012 about 1056 species including varieties have been introduced (Debnath *et al.*, 2014). After the Rio Convention in 1992 the trend of introducing RET species in AJCBIBG got the real momentum. Throughout the year selective RET species being collected from the wild and introduced and

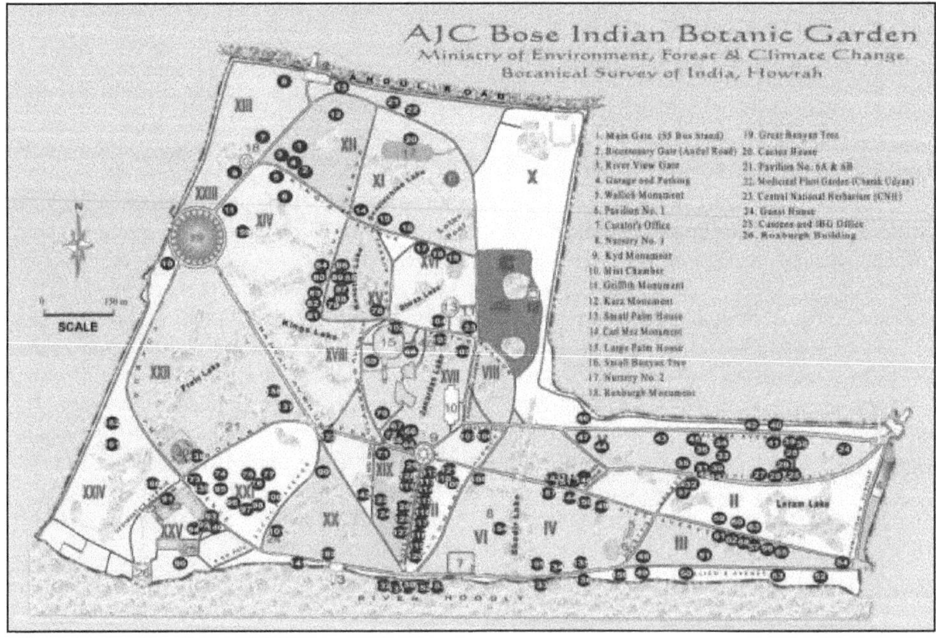

multiplied for the purpose of effective conservation. This garden is divided into 25 divisions representing different phytogeographical regions of India and flora of the world (Chowdhery and Panday, 2007). Selection of plants for introduction in these divisions is also done in accordance to their scheduled zonal preferences. But in due course of time, it is observed that some of these plants are being gradually replaced by different species of so-called Primary Hemi-epiphytes (PHE) while a good numbers are in different stages of grasp.

In nature, true epiphytes or holo-epiphytes root and perch on other plants but never root in soil and complete its entire life cycle anchored to the host plant receiving nutrients only from non-terrestrial sources while a hemi-epiphyte is strictly epiphytic in initial stage of life but becomes rooted in the soil during later stage. Schimper (1903) define the term more precisely by adding a temporal component with the epiphyte. According to him 'hemiepiphytes' are structurally dependent plants that share germination in tree crowns as epiphytes but later on establish contact with the ground via aerial roots that adhere to the host plant and grow downward from the canopy. There are approximately 20 families with hemiephytic plants, mention may be made some of these like, Moraceae, Clusiaceae, Cactaceae, Araceae, *etc*. Hemiepiphytes are again divided into 'strangler' and 'non strangler' category depending upon their capability of self sustaining in due course of development. Puitz and Halbrook (1986) defined the Primary Hemiephiphyte (PHE) as those epiphytes that start their life in trees later establish themselves on the ground as independent trees as mostly found in strangler fig group (*Ficus*) whereas a Secondary Hemiepiphyte (SHE) begins life rooted in soil and later assumes an epiphytic life. Primary Hemiephiphytes are very common and unique forest component in tropics mainly in rain forests. But their occurrence, especially that of 'strangler' category, in a botanic garden may be proved to be fetal for the conservation of rare and endemic plants. Eventually in a botanic garden exclusively designated for *ex-situ* conservation, each and every introduced plant is important from either point of view. In the course of detailed survey, it is found the way PHE are establishing themselves desperately on the trees (Phorophytes) and nibbling them, is a matter of great concern for the conservation practices and at times even identity of the Phorophytes introduced for conservation. The age old Great Banyan tree in AJCBIBG is representing the climax replacement of phorophyte, one poor Date palm (*Borasus flavelifer*) (Chowdhery *et al.*, 2005).

Present work dealt with 163 such cases in AJCBIBG ranging from the new intruders to well established primary hemi-epiphytic *Ficus* trees of family Moraceae and highlight the way the host plants are strangulated and finally swallowed up in due course of time.

Methodology

A thorough survey is done in all the divisions of AJCBIBG from 2014 to 2015 and all such cases are recorded in the field. GPS is used to pinpoint the location of PHE and accordingly plotted on the map. Other field data such as circumference of tree trunk, number of prop roots clinging from the phorophyte stem, circumference of thickest prop root at breast height are recorded. Plant samples are collected to

Table 3.1: Distribution of Hemi-epiphytes on different Phorophytes in AJC Bose Indian Botanic Garden

A	*B*	*C*	*D*	*E*	*F*	*G*	*H*	*I*	*J*
1	*Lagerstroemia parviflora* Roxb. Fam.: Lythraceae	(i)	XII	22°33'44.27"N 88°17'18.14"E	110	0.2	0.20	1100	III
2	*Polyalthia longifolia* (Sonn.) Thwaites Fam.: Annonaceae	(ii)	XII	22°33'44.90"N 88°17'18.10"E	80	0.3	1.2	2000	III
3	*Polyalthia longifolia* (Sonn.) Thwaites Fam.: Annonaceae	(ii)	XII	22°33'43.16"N 88°17'17.38"E	2	0.02	0.4	3	I
4	*Polyalthia longifolia* (Sonn.) Thwaites Fam.: Annonaceae	(iii)	XII	22°33'44.3"N 88°17'17.9"E	1	0.05	1.5	3	I
5	*Polyalthia longifolia* (Sonn.) Thwaites Fam.: Annonaceae	(i)	XIV	22°33'43.6"N 88°17'17.0"E	1	0.06	1.5	4	I
6	*Syzygium samarangense* (Blume) Merr. and L.M.Perry Fam.: Myrtaceae	(i)	XIV	22°33'41.73"N 88°17'18.58"E	20	0.3	1.5	400	I
7	*Dalbergia lanceolaria* L.f. Fam.: Fabaceae	(iii), (iv) and (v)	XIII	22°33'44.13"N 88°17'17.13"E	10	0.25	4	62.5	I
8	*Syzygium cumini* (L.) Skeels Fam.: Myrtaceae	(iv)	XIII	22°33'46.42"N 88°17'17.41"E	30	0.16	1.8	267	I
9	*Carallia brachiata* (Lour.) Merr. Fam.: Rhizophoraceae	(ii)	XIII	22°33'43.14"N 88°17'15.71"E	20	0.10	1.5	133	I
10	*Adenanthera pavonina* L. Fam.: Mimosaceae	(i)	XXIV	22°33'37.62"N 88°17'09.82"E	30	0.3	2	450	I
11	*Melaleuca leucadendra* (L.) L. Fam.: Myrtaceae	(i)	XIV	22°33'40.89"N 88°17'14.52"E	5	0.25	0.8	156	I

*A	B	C	D	E	F	G	H	I	J
12	*Syzygium cumini* (L.) Skeels Fam.: Myrtaceae	(ii)	XII	22°33'47.70"N 88°17'22.93"E	60	0.6	3.5	1029	III
13	*Albizia chinensis* (Osbeck) Merr. Fam.: Mimosaceae	(iv)	X	22°33'47.83"N 88°17'23.45"E	15	0.35	2.5	210	I
14	Dalbergia latifolia Roxb. Fam.: Fabaceae	(iii) and (iv)	XI	22°33'41.77"N 88°17'23.34"E	40	0.35	2	700	II
15	*Cinnamomum camphora* (L.) J.Presl Fam.: Lauraceae	(v)	XI	22°33'41.32"N 88°17'24.31"E	10	0.15	1.5	100	I
16	*Dalbergia sissoo* DC. Fam.: Fabaceae	(ii) and (iii)	XI	22°33'40.90"N 88°17'24.96"E	10	0.25	2	125	I
17	*Pterocarpus indicus* Willd. Fam.: Fabaceae	(i)	XVI	22°33'39.84"N 88°17'25.43"E	10	0.3	2	150	I
18	*Pterocarpus acerifolius* Willd. Fam.: Sterculiaceae	(iii)	XVI	22°33'39.58"N 88°17'25.95"E	20	0.35	1.8	389	I
19	*Berrya cordifolia* (Willd.) Burret Fam.: Malvaceae	(ii) and (iii)	XVI	22°33'39.11"N 88°17'26.73"E	60	0.4	2	1200	II
20	*Garuga pinnata* Roxb. Fam.: Burseraceae	(ii)	XI	22°33'44.86"N 88°17'25.70"E	25	0.10	1.5	167	I
21	*Swietenia mahagoni* (L.) Jacq. Fam.: Meliaceae	(ii)	X	22°33'47.19"N 88°17'25.49"E	10	0.2	2	100	I
22	*Aegle marmelos* (L.) Corrêa Fam.: Rutaceae	(iv)	X	22°33'36.56"N 88°17'26.30"E	2	0.15	1	30	I
23	*Albizia richardiana* (Voigt) King and Prain Fam.: Mimosaceae	(iii)	XVI	22°33'35.52"N 88°17'32.38"E	6	0.50	1.8	167	I

A	*B*	*C*	*D*	*E*	*F*	*G*	*H*	*I*	*J*
24	*Sindora wallichii* Benth. Fam.: Fabaceae	(i)	I	22°33'24.43"N 88°17'59.19"E	2	0.13	1.5	17	I
25	*Bischofia javanica* Blume Fam.: Euphorbiaceae	(ii)	I	22°33'23.71"N 88°17'54.63"E	60	0.17	1.2	850	II
26	*B. javanica* Blume Fam.: Euphorbiaceae	(iv)	I	22°33'23.46"N 88°17'54.17"E	60	0.4	2	1200	III
27	*Casuarina equisetifolia* L. Fam.: Casuarinaceae	(i)	XX	22°33'18.30"N 88°17'24.19"E	30	0.6	1.5	1200	III
28	*Sindora wallichii* Benth. Fam.: Fabaceae	(iv)	I	22°33'24.77"N 88°17'57.49"E	100	0.3	3	1000	II
29	*Syzygium cumini* (L.) Skeels Fam.: Myrtaceae	(iii)	I	22°33'22.84"N 88°17'54.03"E	20	0.25	1.8	278	I
30	*Adenanthera pavonina* L. Fam.: Mimosaceae	(iii)	I	22°33'23.11"N 88°17'53.49"E	50	0.02	1	100	I
31	*Adenanthera pavonina* L. Fam.: Mimosaceae	(i)	I	22°33'23.23"N 88°17'53.31"E	7	0.21	0.7	210	I
32	*Casuarina equisetifolia* L. Fam.: Casuarinaceae	(i)	II	22°33'21.65"N 88°17'52.54"E	30	0.21	2	315	I
33	*Casuarina equisetifolia* L. Fam.: Casuarinaceae	(i)	I	22°33'24.23"N 88°17'50.63"E	30	0.21	1.8	350	I
34	*Casuarina equisetifolia* L. Fam.: Casuarinaceae	(i)	I	22°33'24.60"N 88°17'50.74"E	35	0.25	1.8	486	I
35	*Adenanthera pavonina* L. Fam.: Mimosaceae	(iv)	I	22°33'23.27"N 88°17'49.73"E	40	0.15	1.8	333	I

*A	B	C	D	E	F	G	H	I	J
36	*Dalbergia lanceolaria* L.f. Fam.: Fabaceae	(iv)	I	22°33'25.24"N 88°17'47.91"E	10	0.2	1.5	133	I
37	*Casuarina equisetifolia* L. Fam.: Casuarinaceae	(vi)	II	22°33'22.02"N 88°17'51.24"E	1	0.06	1.5	4	I
38	*Phoenix sylvestris* (L.) Roxb. Fam.: Arecaceae	(iv)	I	22°33'24.64"N 88°17'51.94"E	10	0.2	0.75	267	I
39	*Casuarina equisetifolia* L. Fam.: Casuarinaceae	(iv)	I	22°33'24.92"N 88°17'43.82"E	50	0.2	1.2	833	II
40	*Myroxylon balsamum* var. *pereirae* (Royle) Harms Fam.: Fabaceae	(iii) and (vi)	I	22°33'25.89"N 88°17'53.82"E	30	0.25	1.5	500	I
41	*Schleichera oleosa* (Lour.) Merr. Fam.: Sapindaceae	(vi)	I	22°33'25.71"N 88°17'48.15"E	2	0.07	1.5	9	I
42	*Dalbergia cultrata* Benth. Fam.: Fabaceae	(iv)	I	22°33'26.32"N 88°17'46.49"E	40	0.2	1.8	444	I
43	*Hyphaene coriacea* Gaertn. Fam.: Arecaceae	( ) and (iv)	V	22°33'25.46"N 88°17'44.22"E	10	0.15	0.6	250	I
44	*Hyphaene coriacea* Gaertn. Fam.: Arecaceae	( ) and (iii)	V	22°33'24.49"N 88°17'42.91"E	4	0.2	0.8	100	I
45	*Borassus flabellifer* L. Fam.: Arecaceae	(vi)	I	22°33'25.95"N 88°17'41.99"E	10	0.4	1	400	I
46	*Borassus flabellifer* L. Fam.: Arecaceae	(i)	V	22°33'26.53"N 88°17'41.67"E	30	0.4	1.2	1000	II
47	*Roystonea regia* (Kunth) O.F.Cook Fam.: Arecaceae	(iv)	V	22°33'26.07"N 88°17'40.09"E	100	0.5	1.5	3333	IV

*A	B	C	D	E	F	G	H	I	J
48	*Swietenia mahagoni* (L.) Jacq. Fam.: Meliaceae	(iv)	III	22°33'17.16"N 88°17'44.88"E	50	0.4	1.8	1111	III
49	*Tectona grandis* L.f. Fam.: Verbenaceae	(i)	III	22°33'15.79"N 88°17'45.16"E	2	0.05	1.5	7	I
50	*Casuarina equisetifolia* L. Fam.: Casuarinaceae	(ii)	III	22°33'15.58"N 88°17'47.57"E	100	0.4	1.8	2222	IV
51	*Terminalia catappa* L. Fam.: Combretaceae	(i)	III	22°33'16.04"N 88°17'48.01"E	40	0.35	1.8	778	II
52	*Terminalia catappa* L. Fam.: Combretaceae	(ii)	III	22°33'14.12"N 88°17'54.98"E	100	0.18	2	900	II
53	*Casuarina equisetifolia* L. Fam.: Casuarinaceae	(iv)	III	22°33'14.78"N 88°17'53.08"E	40	0.4	1.2	1333	III
54	*Swietenia macrophylla* King Fam.: Meliaceae	(i) and (iv)	III	22°33'15.19"N 88°17'58.96"E	10	0.1	1.1	91	I
55	*Casuarina equisetifolia* L. Fam.: Casuarinaceae	(iv)	III	22°33'17.40"N 88°17'56.84"E	25	0.3	1.5	500	I
56	*Swietenia macrophylla* King Fam.: Meliaceae	(ii)	III	22°33'17.40"N 88°17'56.62"E	1	0.04	1.8	2	I
57	*Hyphaene thebaica* (L.) Mart. Fam.: Arecaceae	(ii)	III	22°33'17.40"N 88°17'55.88"E	1	0.04	0.9	4	I
58	*Phoenix dactylifera* L. Fam.: Arecaceae	(v)	III	22°33'17.56"N 88°17'55.38"E	30	0.15	0.8	562	II
59	*Casuarina equisetifolia* L. Fam.: Casuarinaceae	(v)	II	22°33'17.82"N 88°17'54.76"E	50	0.3	1.8	833	II

**A*	*B*	*C*	*D*	*E*	*F*	*G*	*H*	*I*	*J*
60	*Swietenia macrophylla* King Fam.: Meliaceae	(ii)	II	22°33'19.75"N 88°17'51.01"E	50	0.35	1.5	1167	III
61	*Swietenia macrophylla* King Fam.: Meliaceae	(i)	III	22°33'19.85"N 88°17'51.98"E	10	0.4	1.5	267	I
62	*Swietenia macrophylla* King Fam.: Meliaceae	(i)	III	22°33'18.65"N 88°17'52.27"E	3	0.10	1.5	20	I
63	*Swietenia mahagoni* (L.) Jacq. Fam.: Meliaceae	(ii)	II	22°33'19.12"N 88°17'53.06"E	1	0.07	1.5	5	I
64	*Mitragyna parvifolia* (Roxb.) Korth. Fam.: Rubiaceae	(ii)	XVII	22°33'25.45"N 88°17'27.04"E	1	0.05	1.5	3	I
65	*Pterocarpus indicus* Willd. Fam.: Fabaceae	(iii)	XVII	22°33'25.66"N 88°17'26.04"E	60	0.2	1.5	800	II
66	*Haldina cordifolia* (Roxb.) Ridsdale Fam.: Rubiaceae	(i)	XVII	22°33'25.87"N 88°17'26.17"E	28	0.6	2	840	II
67	*Madhuca longifolia* var. *latifolia* (Roxb.) A.Chev. Fam.: Sapotaceae	(ii)	XVII	22°33'26.10"N 88°17'25.46"E	1	0.03	1.2	2.5	I
68	*Harpullia cupanioides* Roxb. Fam.: Sapindaceae	(ii)	XVII	22°33'29.02"N 88°17'28.20"E	60	0.6	3	1200	III
69	*Swietenia mahagoni* (L.) Jacq. Fam.: Meliaceae	(i)	XVII	22°33'30.33"N 88°17'24.12"E	1	0.07	1.5	5	I
70	*Schleichera oleosa* (Lour.) Merr. Fam.: Sapindaceae	(ii)	XVII	22°33'28.21"N 88°17'23.38"E	50	0.15	3	250	I
71	*Swietenia mahagoni* (L.) Jacq. Fam.: Meliaceae	(i)	X	22°33'26.44"N 88°17'23.10"E	50	0.12	1.8	333	I

A	*B*	*C*	*D*	*E*	*F*	*G*	*H*	*I*	*J*
72	*Swietenia mahagoni* (L.) Jacq. Fam.: Meliaceae	(ii)	XVII	22°33'26.65"N 88°17'23.25"E	10	0.05	1.8	28	I
73	*Swietenia mahagoni* (L.) Jacq. Fam.: Meliaceae	(ii)	XXI	22°33'24.34"N 88°17'10.69"E	1	0.05	1.2	4	I
74	*Filicium decipiens* (Wight and Arn.) Thwaites Fam.: Sapindaceae	(i)	XXI	22°33'26.41"N 88°17'11.67"E	2	0.2	1.5	27	I
75	*Dolichandrone spathacea* (L.f.) Seem. Fam.: Bignoniaceae	(i) and (ii)	XXI	22°33'24.49"N 88°17'13.19"E	30	0.3	4	225	I
76	*Kigelia africana* (Lam.) Benth. Fam.: Bignoniaceae	(i)	XXI	22°33'25.06"N 88°17'14.20"E	10	0.3	3	100	I
77	*Kigelia africana* (Lam.) Benth. Fam.: Bignoniaceae	(v)	XXI	22°33'22.68"N 88°17'15.95"E	5	0.15	1.5	33	I
78	*Bischofia javanica* Blume Fam.: Euphorbiaceae	(iv)	XV	22°33'33.66"N 88°17'24.37"E	50	0.4	3	667	II
79	*Swietenia macrophylla* King Fam.: Meliaceae	(iii)	XV	22°33'34.15"N 88°17'21.38"E	2	0.04	1.2	7	I
80	*Adenanthera pavonina* L. Fam.: Mimosaceae	(iv)	XIV	22°33'34.73"N 88°17'20.88"E	10	0.12	1	120	I
81	*Albizia lebbeck* (L.) Benth. Fam.: Mimosaceae	(i)	XIV	22°33'34.27"N 88°17'20.45"E	4	0.05	1	20	I
82	*Albizia lebbeck* (L.) Benth. Fam.: Mimosaceae	(i)	XIV	22°33'34.53"N 88°17'20.37"E	1	0.05	0.8	6	I
83	*Albizia lebbeck* (L.) Benth. Fam.: Mimosaceae	(iv)	XIV	22°33'34.37"N 88°17'20.80"E	0	0	1.2	0	I

*A	B	C	D	E	F	G	H	I	J
84	*Cassine glauca* (Rottb.) Kuntze Fam.: Celastraceae	(iv)	XIV	22°33'36.37"N 88°17'21.43"E	1	0.05	1.2	4	I
85	*Dalbergia sissoo* DC. Fam.: Fabaceae	(iv)	XV	22°33'36.19"N 88°17'23.25"E	4	0.05	1.2	4	I
86	*Swietenia macrophylla* King Fam.: Meliaceae	(i)	XV	22°33'37.31"N 88°17'22.34"E	10	0.15	1.2	125	I
87	*Lagerstroemia speciosa* (L.) Pers. Fam.: Lythraceae	(ii)	XV	22°33'35.72"N 88°17'22.32"E	0	0	1.5	0	I
88	*Bischofia javanica* Blume Fam.: Euphorbiaceae	(ii)	XV	22°33'36.08"N 88°17'22.71"E	1	0.04	0.8	5	I
89	*Hevea brasiliensis* (Willd. ex A.Juss.) Müll.Arg. Fam.: Euphorbiaceae	(iii)	X	22°33'45.33"N 88°17'30.90"E	1	0.08	2	4	I
90	*Casuarina equisetifolia* L. Fam.: Casuarinaceae	(i) and (iii)	XXV	22°33'19.21"N 88°17'10.54"E	50	0.15	1	750	II
91	Beyond recognition	(ii)	XXV	22°33'22.31"N 88°17'9.15"E	200	0.5	5	2000	III
92	*Parinarium nitidum* Hook.f. Fam.: Chrysobalanaceae	(i)	XXV	22°33'20.39"N 88°17'11.67"E	20	0.5	1.6	625	II
93	*Syzygium cumini* (L.) Skeels Fam.: Myrtaceae	(ii)	XXV	22°33'21.51"N 88°17'11.88"E	2	0.04	1.5	5	I
94	*Lagerstroemia speciosa* (L.) Pers. Fam.: Lythraceae	(ii)	XXV	22°33'20.51"N 88°17'11.21"E	5	0.1	0.8	63	I
95	Beyond recognition	(ii)	XXI	22°33'22.25"N 88°17'13.81"E	80	0.4	1.5	2133	IV

*A	B	C	D	E	F	G	H	I	J
96	*Syzygium cumini* (L.) Skeels Fam.: Myrtaceae	(i)	XXI	22°33'24.12"N 88°17'15.72"E	15	0.1	0.9	167	I
97	*Strychnos nux-vomica* L. Fam.: Loganiaceae	(i)	XXI	22°33'23.86"N 88°17'17.46"E	1	0.07	0.8	9	I
98	*Hopea odorata* Roxb. Fam.: Dipterocarpaceae	(i)	XXI	22°33'21.87"N 88°17'16.31"E	5	0.7	0.9	389	I
99	*Terminalia arjuna* (Roxb. ex DC.) Wight and Arn. Fam.: Combretaceae	(i)	XX	22°33'26.26"N 88°17'20.15"E	0	0	0.8	0	I
100	*Swietenia macrophylla* King Fam.: Meliaceae	(i)	XXI	22°33'22.19"N 88°17'17.97"E	2	0.04	0.9	9	I
101	*Swietenia macrophylla* King Fam.: Meliaceae	(ii)	XX	22°33'22.20"N 88°17'17.67"E	2	0.04	0.8	10	I
102	*Vitex pinnata* L. Fam.: Verbenaceae	(i)	XVII	22°33'31.03"N 88°17'28.31"E	2	0.06	0.8	15	I
103	*Arfeuillea arborescens* Pierre ex Radlk. Fam.: Sapindaceae	(i)	XVII	22°33'32.33"N 88°17'30.41"E	3	0.08	1.5	16	I
104	*Enterolobium cyclocarpum* (Jacq.) Griseb. Fam.: Mimosaceae	(i)	XVI	22°33'32.55"N 88°17'31.30"E	0	0	2	0	I
105	*Albizia richardiana* (Voigt) King and Prain Fam.: Mimosaceae	(iii)	XI	22°33'45.52"N 88°17'26.62"E	10	0.04	1.2	33	I
106	*Syzygium cumini* (L.) Skeels Fam.: Myrtaceae	(i)	VIII	22°33'26.47"N 88°17'32.00"E	1	0.5	0.6	83	I
107	Beyond recognition	(ii)	VIII	22°33'25.38"N 88°17'30.09"E	150	0.6	3	3000	IV

*A	B	C	D	E	F	G	H	I	J
108	*Vitex peduncularis* Wall. ex Schauer Fam.: Verbenaceae	(iv)	VI	22°33'23.60"N 88°17'32.11"E	2	0.08	0.4	40	I
109	*Dimocarpus longan* Lour. Fam.: Sapindaceae	(ii)	VII	22°33'20.57"N 88°17'29.86"E	28	0.2	0.6	933	II
110	*Swietenia mahagoni* (L.) Jacq. Fam.: Meliaceae	(ii)	VII	22°33'21.32"N 88°17'28.17"E	30	0.06	3	60	I
111	*Swietenia mahagoni* (L.) Jacq. Fam.: Meliaceae	(ii)	XIX	22°33'21.33"N 88°17'27.20"E	25	0.09	2.8	80	I
112	*Swietenia mahagoni* (L.) Jacq. Fam.: Meliaceae	(i)	VII	22°33'20.85"N 88°17'28.03"E	20	0.1	3.1	65	I
113	*Swietenia mahagoni* (L.) Jacq. Fam.: Meliaceae	(ii)	VII	22°33'20.58"N 88°17'27.75"E	28	0.12	3.2	105	I
114	*Swietenia mahagoni* (L.) Jacq. Fam.: Meliaceae	(iv)	VII	22°33'20.05"N 88°17'27.77"E	31	0.15	3.1	150	I
115	*Swietenia mahagoni* (L.) Jacq. Fam.: Meliaceae	(i)	VII	22°33'19.99"N 88°17'27.68"E	40	0.2	3.5	229	I
116	*Swietenia mahagoni* (L.) Jacq. Fam.: Meliaceae	(i)	VII	22°33'19.92"N 88°17'27.76"E	25	0.07	2.9	60	I
117	*Swietenia mahagoni* (L.) Jacq. Fam.: Meliaceae	(i)	VII	22°33'19.41"N 88°17'27.54"E	21	0.07	3	49	I
118	*Swietenia mahagoni* (L.) Jacq. Fam.: Meliaceae	(i)	VII	22°33'19.15"N 88°17'27.51"E	40	0.05	2.9	69	I
119	*Swietenia mahagoni* (L.) Jacq. Fam.: Meliaceae	(iv)	VII	22°33'18.09"N 88°17'27.17"E	32	0.11	3.1	114	I

*A	B	C	D	E	F	G	H	I	J
120	*Swietenia mahagoni* (L.) Jacq. Fam.: Meliaceae	(ii)	VII	22°33'23.67"N 88°17'28.57"E	45	0.12	3	180	I
121	*Ehretia laevis* Roxb. Fam.: Boraginaceae	(i)	VII	22°33'24.02"N 88°17'29.25"E	1	0.06	0.8	6	I
122	*Calophyllum inophyllum* L. Fam.: Cluciaceae	(ii)	XIX	22°33'28.29"N 88°17'27.41"E	1	0.03	1.2	3	I
123	*Swietenia mahagoni* (L.) Jacq. Fam.: Meliaceae	(i)	XIX	22°33'30.33"N 88°17'24.12"E	1	0.05	2	3	I
124	*Swietenia mahagoni* (L.) Jacq. Fam.: Meliaceae	(i)	XIX	22°33'30.51"N 88°17'25.96"E	5	0.15	2.5	194	I
125	*Swietenia mahagoni* (L.) Jacq. Fam.: Meliaceae	(iii)	XIX	22°33'30.62"N 88°17'25.61"E	2	0.02	1.8	2	I
126	*Swietenia mahagoni* (L.) Jacq. Fam.: Meliaceae	(i)	XIX	22°33'29.22"N 88°17'23.98"E	1	0.04	1.6	3	I
127	*Swietenia mahagoni* (L.) Jacq. Fam.: Meliaceae	(iv)	XIX	22°33'29.10"N 88°17'23.13"E	1	0.03	2	2	I
128	*Swietenia mahagoni* (L.) Jacq. Fam.: Meliaceae	(iv)	XIX	22°33'27.26"N 88°17'23.01"E	1	0.03	0.9	3	I
129	*Swietenia mahagoni* (L.) Jacq. Fam.: Meliaceae	(i)	XIX	22°33'24.89"N 88°18'0.63"E	40	0.12	1	480	I
130	*Tectona grandis* L.f. Fam.: Verbenaceae	(ii)	VII	22°33'16.95"N 88°17'27.89"E	5	0.04	0.8	25	I
131	*Tectona grandis* L.f. Fam.: Verbenaceae	(i)	VII	22°33'17.11"N 88°17'27.38"E	2	0.10	1.1	18	I

*A	B	C	D	E	F	G	H	I	J
132	*Tectona grandis* L.f. Fam.: Verbenaceae	(iii)	VII	22°33'17.05"N 88°17'26.53"E	5	0.2	2	50	I
133	*Casuarina equisetifolia* L. Fam.: Casuarinaceae	(ii)	VII	22°33'16.68"N 88°17'37.07"E	35	0.3	1.1	955	II
134	*Shorea robusta* Gaertn. Fam.: Dipterocarpaceae	(iii)	IV	22°33'17.63"N 88°17'38.32"E	2	0.06	1	12	I
135	*Cratoxylum cochinchinense* (Lour.) Blume Fam.: Hypericaceae	(iii)	IV	22°33'17.42"N 88°17'39.24"E	2	0.04	0.8	10	I
136	*Casuarina equisetifolia* L. Fam.: Casuarinaceae	(i)	IV	22°33'16.22"N 88°17'40.08"E	36	0.07	0.9	280	I
137	*Nauclea orientalis* (L.) L. Fam.: Rubiaceae	(i)	XXII	22°33'27.74"N 88°17'17.91"E	0	0	1.2	0	I
138	*Mitragyna parvifolia* (Roxb.) Korth. Fam.: Rubiaceae	(iii)	XXII	22°33'25.01"N 88°17'10.52"E	10	0.01	2	5	I
139	*Labramia bojeri* A.DC. Fam.: Sapotaceae	(ii)	XXI	22°33'22.51"N 88°17'12.14"E	2	0.04	0.8	10	I
140	*Toona ciliata* M.Roem. Fam.: Meliaceae	(i)	XXV	22°33'20.56"N 88°17'12.37"E	2	0.08	0.8	20	I
141	*Terminalia catappa* L. Fam.: Combretaceae	(i)	XX	22°33'18.19"N 88°17'18.82"E	0	0	2	0	I
142	*Borassus flabellifer* L. Fam.: Arecaceae	(vi)	XIX	22°33'21.30"N 88°17'24.51"E	15	0.15	1.1	205	I
143	*Polyalthia longifolia* (Sonn.) Thwaites Fam.: Annonaceae	(i)	V	22°33'22.67"N 88°17'37.71"E	40	0.6	1.4	1714	III

*A	B	C	D	E	F	G	H	I	J
144	*Polyalthia longifolia* (Sonn.) Thwaites Fam.: Annonaceae	(iv)	IV	22°33'21.59"N 88°17'39.50"E	2	0.12	2	12	I
145	*Casuarina equisetifolia* L. Fam.: Casuarinaceae	(iii)	X	22°33'38.72"N 88°17'33.56"E	0	0	1.8	0	I
146	*Swietenia mahagoni* (L.) Jacq. Fam.: Meliaceae	(i)	XI	22°33'47.14"N 88°17'26.99"E	1	0.18	1.5	12	I
147	*Casuarina equisetifolia* L. Fam.: Casuarinaceae	(i)	I	22°33'22.33"N 88°17'49.52"E	1	0.03	0.9	3	I
148	*Casuarina equisetifolia* L. Fam.: Casuarinaceae	(iii)	I	22°33'23.27"N 88°17'48.62"E	1	0.04	0.9	4	I
149	*Dalbergia latifolia* Roxb. Fam.: Fabaceae	(iii)	IV	22°33'21.14"N 88°17'42.81"E	50	0.3	2	750	II
150	*Attalea speciosa* Mart. Fam.: Arecaceae	(i)	IV	22°33'21.17"N 88°17'41.71"E	150	0.9	4	3375	IV
151	*Attalea speciosa* Mart. Fam.: Arecaceae	(ii)	V	22°33'22.64"N 88°17'40.14"E	40	0.7	0.9	3111	IV
152	*Tectona grandis* L.f. Fam.: Verbenaceae	(ii)	VI	22°33'17.46"N 88°17'29.07"E	2	0.02	1.2	3	I
153	*Peltophorum pterocarpum* (DC.) K.Heyne Fam.: Caesalpinaceae	(iii)	VI	22°33'17.22"N 88°17'29.67"E	0	0	1.2	0	I
154	*Mimusops elengi* L. Fam.: Sapotaceae	(iv)	VI	22°33'19.22"N 88°17'32.66"E	130	0.10	1.8	722	II
155	*Litchi chinensis* Sonn. Fam.: Sapindaceae	(iii)	X	22°33'45.44"N 88°17'34.19"E	10	0.5	1.5	333	II

*A	B	C	D	E	F	G	H	I	J
156	*Casuarina equisetifolia* L. Fam.: Casuarinaceae	(i)	III	22°33'16.70"N 88°17'40.24"E	1	0.02	1.8	1	I
157	*Mimusops elengi* L. Fam.: Sapotaceae	(vii)	X	22°33'44.37"N 88°17'33.83"E	20	0.20	1.8	222	I
158	*Terminalia arjuna* (Roxb. ex DC.) Wight and Arn. Fam.: Combretaceae	(i)	XXII	22°33'27.82"N 88°17'18.00"E	1	0.35	1.2	29	I
159	*Terminalia arjuna* (Roxb. ex DC.) Wight and Arn. Fam.: Combretaceae	(i)	XIV	22°33'37.33"N 88°17'16.56"E	2	0.20	1.2	33	I
160	*Diospyros mespiliformis* Hochst. ex A.DC. Fam.: Ebenaceae	(i)	XXIV	22°33'23.72"N 88°17'07.56"E	1	0.05	1.2	4	I
161	*Swietenia mahagoni* (L.) Jacq. Fam.: Meliaceae	(iii)	XXIV	22°33'25.74"N 88°17'04.33"E	2	0.06	1.8	7	I
162	*Dalbergia lanceolaria* L.f. Fam.: Fabaceae	(i)	XXIV	22°33'27.20"N 88°17'05.09"E	2	0.05	1.5	7	I
163	*Coccoloba uvifera* L. Fam.: Polygonaceae	(i)	XX	22°33'44.08"N 88°17'34.34"E	2	0.25	1.5	33	I

Table Legends: **A.** Sl. No.; **B.** Phorophyte; **C.** Primary Hemi-epiphyte; **D.** Garden Division; **E.** Latitude and Longitude; **F.** No. of prop-roots of PHE at Breast Height; **G.** Circumference of thickest prop-root of PHE (in meter); **H.** Total Circumference of the stem at breast height (in meter); **I.** Index of PHE infestation; **J.** Category. ***Primary Hemi-epiphytes***: **(i)** *Ficus benghalensis* L.; **(ii)** *F. benjamina* L.; **(iii)** *F. religiosa* L.; **(iv)** *F. rumphii* Blume; **(v)** *F. virens* Aiton; **(vi)** *F. retusa* L.; **(vii)** *F. racemosa* L.

identify the PHE and phorophytes. Photographs from different angles are also taken for record. As the determination of age either of PHE or phorophyte is not feasible, it is assumed that the thickness and number of prop-roots is directly proportional to the age of PHE in the ideal natural conditions. As the extant of establishment of PHE on phorophyte is directly proportional to the number of prop-roots and circumference of prop-roots and indirectly proportional to the circumference of the tree trunk, the index of PHE infestation is deduced from a simple product of number of prop-roots and circumference of the thickest prop-root multiplied by 100 and divided by the circumference of trunk of phorophyte along with prop-roots at breast height. Increased value of root number or circumference increases the extent of PHE infestation. While thick girth of phorophyte signifies their capability to pacify the effect of PHE. The field data are presented in the form of Tables 3.1–3.3. Therefore,

$$\text{Index of PHE infestation} = \frac{\text{No. of prop - roots} \times \text{Circumference of thickest prop - root}}{\text{Cumulative Circumference of phorophyte} + \text{prop - roots}} \times 100$$

To understand the different cases of occurrence of PHE in the botanic garden, all the available data is divided into four categories: Category I: New comer PHE (Index of PHE infestation ≤ 500); Category II: Full grown PHE where identity of the phorophyte is suppressed by the PHE (Index of PHE infestation ≤ 1000); Category III: Full grown PHE where identity of the phorophyte is not clear (Index of PHE infestation ≤ 2000); Category IV: PHE almost replaces the phorophytes (Index of PHE infestation ≤ 4000).

Table 3.2: 10 Extreme Cases of PHE

Sl.No.	*Sl. No. in the PHE Table*	*Co-efficient Percentage*
1	150	3375
2	47	3333
3	151	3111
4	107	3000
5	50	2222
6	95	2133
7	91	2000
8	2	2000
9	143	1714
10	53	1333

Result and Discussion

The genus *Ficus* in the family Moraceae has approximately 750 species worldwide and is mostly distributed in tropics and subtropics (Frank *et al.*, 1992). 89 *Ficus* spp. occur in India (Chaudhary and Panday, 2007) out of which 36 species occur in AJCBIBG (Debnath *et al.*, 2014). Only 7 species are found to act as PHE in the garden (Chart 3.1).

A big strangler fig, with its intertwined aerial-roots wrapping around and encasing the remains of its one-time host, is an extraordinary sight (Plate 3.1:A-E). The stranglers, however, produce anastomosing aerial-roots that eventually enclose and kill the host, leaving the fig free-standing. While coming across the PHE infested trees of AJCBIBG, authors met with several such conditions as depicted in serial number 91, 95 and 107 of Table 3.1, where even the identity of host phorophyte was not clear (Plate 3.2; D). At several other instances, the extent of PHE infestation was very high (Table 3.2) and phorophyte was represented only by few branches protruding at the crown level.

Table 3.3 represents the frequency of 10 most susceptible phorophytes. Result shows that some plants seem to be more susceptible to epiphytes than others. Plant morphology plays a significant role in epiphyte and hemi-epiphyte recruitment and establishment (Male and Roberts, 2005). One feature that could facilitate PHE establishment could be epidermal texture and roughness. For example, the smooth, hard outer surface of *Delonix regia* and *Roystonea regia* seems to discourage the accumulation of detritus. It is the detritus material which is needed for epiphyte seeds to take hold and germinate. Plants like *Swietenia mahagoni, S. macrophylla, Casuarina equisetifolia, Syzygium cumini, Polyalthia longifolia* provides ample scope in terms of rough surface and craters formed in the old stem to settle the seeds of

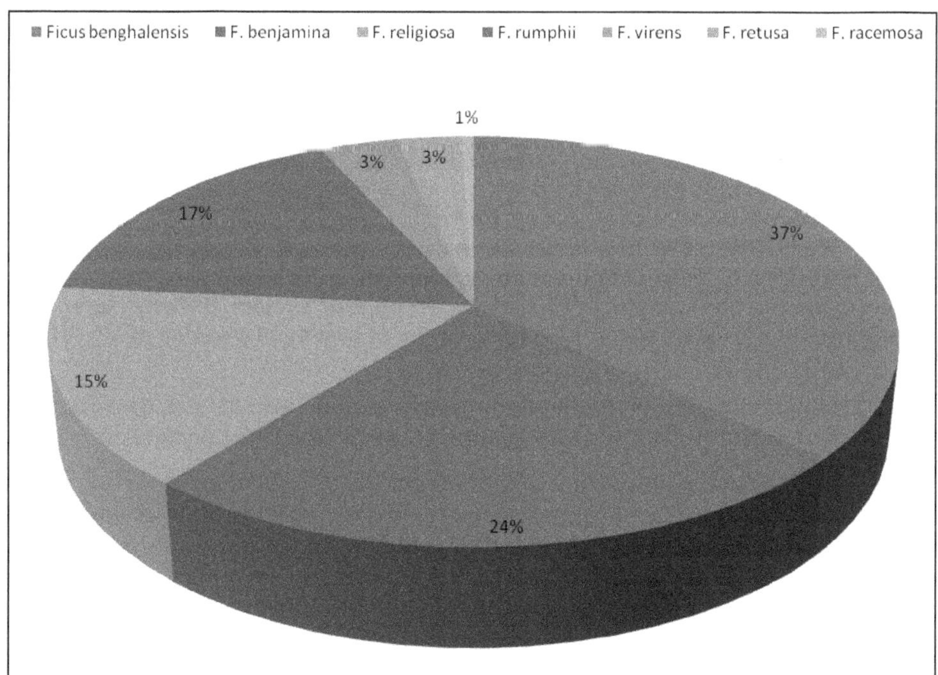

Chart 3.1: Percentage Distribution of different PHE Species.

Figure 3.1

A. PHE establishes itself in a crater on the palm trunk, B. Single tap root coming out of the crater, C. Mess of tap root system intermingling together to form a deadly cage around the phorophyte, D. Thickening ofPHE root system, E. Remains of dead phorophyte in a hollow cage, F. Displaced centre of gravity of phorophyte by PHE.

PHE. Through competition for sunlight, nutrients and vascular constriction of the phloem and xylem tissues, the host plant is overwhelmed and perishes.

The fact that trees supporting a hemi-epiphytic fig were more likely to fall than those that did not have a fig is interesting. It indicates that the hemi-epiphytic figs may be weakening their hosts in some way, although obviously this is not to the selective advantage of the fig. Increased wind resistance in the canopy would appear to be the most likely cause, although other factors such as weakened root systems through competition for resources or asymmetric loading of the host trunk through unequal growth of the fig crown may play a part (Plate 3.1: F). It is also possible that hemi-epiphytic figs preferentially colonize overly mature trees, if these offer more

Figure 3.2

A. *Ficus religiosa* totally covered a *Roystonea regia* plant, B. Dead remains of *R. regia* protruding from the top, C. PHE along with other lianas intermingle a phorophyte, D. Roots of a PHE passing through the dead stem of a unrecognizable phorophyte, E. Cage like structured formed after the death and decay of phorophyte, F. PHE engulfing a concrete structure (Plant Gabion), G. An old building of AJCBIBG affected by PHE.

suitable micro-sites for seedling establishment, such as notches with rotten wood or more open crowns as a result of fallen branches (Plate 3.1; A).

Table 3.3: Frequency of 10 Most Susceptible Phorophytes

Sl.No.	*Name*	*Family*	*No. of occurrence*
1	*Swietenia mahagoni* (L.) Jacq.	Meliaceae	25
2	*Casuarina equisetifolia* L.	Casuarinaceae	18
3	*Swietenia macrophylla* King	Meliaceae	9
4	*Syzygium cumini* (L.) Skeels	Myrtaceae	6
5	*Polyalthia longifolia* (Sonn.) Thwaites	Annonaceae	6
6	*Adenanthera pavonina* L.	Mimosaceae	5
7	*Tectona grandis* L.f.	Verbenaceae	5
8	*Bischofia javanica* Blume	Euphorbiaceae	4
9	*Albizia lebbeck* (L.) Benth.	Mimosaceae	3
10	*Terminalia arjuna* (Roxb. ex DC.) Wight and Arn.	Combretaceae	3

Damage to the concrete structure by PHE is also an important matter of human concern that destroys old buildings and other structures. In the botanic garden too, these PHE has engulfed fully or partially many important building of British period that has heritage importance (Plate 3.2; F-G).

There are several methods for PHE dispersal and subsequent recruitment into susceptible host species. *Ficus* fruit is mammalian and avian dispersed. Many animals that are frugivors, after feasting on *Ficus* fruits, will seek shelter for roosting in a thick canopy of leaves such as a palm crown. In AJCBIBG many mammals like Asian Palm Civet, Small Indian Civet, Northern Plain Grey Langur, bats, squirrel and different birds may be the possible carrier of seeds of these figs. (Debnath *et al.*, 2012, 2014)

Though the occurrence of primary hemi-epiphytes is a natural process and death of phorophytes due to these stranglers is also not very uncommon in the tropical rainforests. But for a botanic garden, where plants are conserved, presence of such primary hemi-epiphytes may not be a healthy sign for the long term survival of rare and endemic species. Every care must be taken to remove the PHE at its early stage. For this, a thorough screening of susceptible plants at regular interval is necessary. After this study a task force composed of field staff, supervised by scientific personnel is framed in AJCBIBG who is involved in removing the PHE of Category I. At the same time phorophytes of Category II, III and IV (wherever possible) are labeled and recorded properly with latitude and longitude for future reference and identification. Special care is also given in introducing any new fig species in the garden and are planted is remote, isolated localities so the dispersal of PHE can be controlled to a possible extant.

Acknowledgement

The authors are thankful to the Director, Botanical Survey of India for facilities and encouragement.

References

Chowdhery, H.J and Pandey, D.S, 2007. *Plants of Indian Botanic Garden, Botanical Survey of India,* Botanical Survey of India, Kolkata.

Debnath, H.S., Mahapatra, H.S., Hameed, S.S. and Sreekumar, P.V. 2014. *Census of Plants in AJC Bose Indian Botanic Garden: A Report*. Botanical Survey of India, Kolkata. pp. 60.

Debnath, H.S., Roychowdhury, B. 2012. *Birds, Butterflies and Associate Plants of AJC Bose Indian Botanic Garden*. Botanical Survey of India, Kolkata.

Debnath, H.S., Roychowdhury, B. 2014. *Reptiles, Mammals and Amphibians of AJC Bose Indian Botanic Garden*. Botanical Survey of India, Kolkata.

Frank, R.J., Knight R.J. and Nadel, H. 1992. Escapees and Accomplices: The Naturalization of Exotic Ficus and Their Associated Faunas in Florida. *Florida Entom.* 75(1): 29-38.

Male, T.D. and Roberts, G. E. 2005. Host associations of the strangler fig, Ficus watkinsiana, in a subtropical Queensland rain forest. *Aust. Ecol.* 30:229-236.

Puitz FE and Holbrook NM. 1986. Notes on the natural history of hemiepiphytes. *Selbyana* 9: 61–69.

Schimper AFW. 1903. *Plant geography upon a physiological basis*. Oxford: Clarendon Press.

Chpater 4

Conservation of Plants of different Phyto-geographical Region of India in Botanic Garden of Indian Republic (BGIR), Noida

Sheo Kumar

Botanic Garden of Indian Republic, Botanical Survey of India, Govt. of India, Ministry of Environment, Forest and Climate Change, Capt. Vijyant Thapar Marg, Sector 38A, NOIDA – 201 303, U.P.
E-mail: drsheo_kumar@rediffmail.com

Abstract

*As per International Agenda/recommendations of the Conferences of UNO, IUCN, CBD and requirement of Global Mission, 'Botanic Garden of Indian Republic (BGIR)' was established as 'National Botanic Garden' by the Ministry of Environment, Forest and Climate Change in Delhi NCR (NOIDA) and is spread over on 163.79 acre land area. This garden is dedicatedly engaged in conservation of endemic and threatened plants of different phyto-geographical regions of the country under ex-situ conservation along with other components like research, training, environmental education/awareness and recreation. Though, presently BGIR is in nascent stage, even then, more than 10,500 (10,666) individuals of 876 plant species (nearly 12 individual of each species) among 37 species listed in Red Data Book, brought from 23 States of the country are conserved under **ex-situ** conservation in specified 5 Nurseries, 3 Arboretum, Thematic Gardens, Cactus and Succulents, Medicinal Plant Garden, Landscaped Horticulture Garden and Water body. Thus, this garden* *is functioning as a national repository of living plants. Besides, out of listed 42 Articles of CBD (1992), BGIR is complying altogether 10 Articles (among 7 Articles are directly i.e. Article 7, 9, 12, 13, 15, 17 and 18 and 3 Articles are indirectly i.e. Article 6, 8 and 10). In continuation, efforts taken/under process will entail BGIR to achieve further requirements in near future in national interest as committed during Earth Summit held at Rio de Janeiro, Brazil in 1992.*

Keywords: *Global mission, Endemic and threatened plants, Ex-situ conservation Phyto-geographical.*

INTRODUCTION

The Republic of India, in hindi *Bhārat Gaṇarājya* (Clémentin-Ojha, **2014;** McWhirter and McWhirter, 1964) is seventh-largest, second-most populous, pluralistic, multilingual, multiethnic societal country with more than 20 official symbols prefix 'National' with entities are intrinsic to the Indian identity and heritage, governed under parliamentary system by consisting 29 States and 7 Union Territories (UT's). It lies within the Indomalaya ecozone as having predominantly four major climatic groupings *i.e.* tropical wet, tropical dry, subtropical humid and montane (*Heitzman and Worden, 1996*), three biodiversity hotspots (*CI, 2007; Roach,* 2005) and is one among 17 megadiverse countries. Thus, it contains huge number of flora and fauna with diversity in varying protected and unprotected habitats which ranges from tropical rainforest of the Andaman Islands, Western Ghats and North-East to coniferous forest of the Himalaya. Between these extremes lie the moist deciduous 'sal' forest of eastern India; the dry deciduous 'teak' forest of central and southern India and the 'babul' dominated thorn forest of the central Deccan and western Gangetic plain (*Tritsch, 2001*). Currently, based on available data on taxa and plant diversity, India is placed at tenth position in the world and fourth in Asia. Keeping in view of inventorisarion and survey in approx. 70 per cent geographical areas, so far, 47,791 species of plants (Algae, Bacteria, Fungi, Bryophytes, Pteridophytes, Gymnosperms and Angiosperms) have been described (Plant Discoveries: BSI, 2015) among many Indian species descend from taxa originating in Gondwana *i.e.* Indian plate separated more than 105 million years ago (*Crame and Owen, 2002*). Subsequent movement of Peninsular India's intended further collision with the Laurasian landmass and set off mass exchange of species. Epochal volcanism and climatic changes taken place 20 million years ago forced a mass extinction (*Karanth, 2006*). Besides, above facts, reduction in number of flora and loss of biodiversity and livelihood of dependent society is in danger and still continuing by obviously different causes *i.e.* faster industrialisation, urbanisation, deforestations and others. As similar situation is prevailing almost in most developed and developing countries, experiencing adverse impacts on their environmental conditions and loss of flora and fauna, biodiversity and thus, looking for devices to mitigate/restore the same by adopting sustainable development process, conservation, awareness and restoration for future generation.

India, being a developing country, participated in various conferences, seminars, meetings and signed various International agreements of United Nations

Organisation (UNO) in 1972, International Union for Conservation of Nature (IUCN) in 1975 and 1985 and the Convention of Biological Diversity (CBD) during Earth Summit in 1992, a 'National Botanic Garden', now known as 'Botanic Garden of Indian Republic' was established on 19.05.1997 over 200 acre land area in Sector – 38A, NOIDA by the Ministry of Environment and Forests (now known as Ministry of Environment, Forest and Climate Change) to fulfil the requirement of agreement, International Agenda and Global Mission through Botanical Survey of India with the following objectives:

- ☆ *Ex-situ* Conservation of endemic and threatened plant species of representative ecosystems of the country.
- ☆ Serve as a 'Centre of Excellence' for research and training.
- ☆ Create public awareness through educational programmes on plant diversity and conservation.
- ☆ Recreation.

Thus, as per definition of botanic garden *i.e. 'an institution holding documented collections of living plants for the purposes of scientific research, conservation, display and education'*, now BGIR is functioning as a living repository at national level for conserving endemic and threatened plant species of different phyto-geographical areas of the country along with other objectives and activities of the garden.

Material and Methods

Botanic Garden of Indian Republic (BGIR) is situated at 28°33′27″ N. lat. to 28°34′00″ N. lat. and 77°19′16″ E. long. to 77°20′13″ E. long. and altitude ranging from 197 to 202 MSL which is located in flood plain of river Yamuna having alluvial soil known as *khaddar*, represents fairly coarse sand with very fine texture on surface soil, greyish-brownish to grey depending upon depth, deficient in nitrogen, humus and phosphoric acid. However, soil contains high percentage of soluble salts of sodium, calcium and magnesium as pH of soil is alkaline and moisture content is very-very low during most part of the year. Before establishment of this garden, land area was under cultivation of cereals [*Pennisetum glaucum* (L.)R.Br. (Bajra), *Sorghum bicolor* (L.) Moench (Jowar), *Zea mays* L. (Maize), *Avena sativa* L. (Oat), *Oryza sativa* L. (Paddy) and *Triticum* aestivum L. (Wheat); pulses [*Cicer arietinum* L. (Gram) and *Pisum sativum* L. (Pea)]; mustard (*Brassica nigra* L.); vegetables [*Solanum melongena* L. (Brinjal), *Brassica oleracea* var. *capitata* (Cabbage), *Brassica oleracea* var. *botrytis* (Cauliflower), *Cucurbita moschata* Duchesne ex Poir. (Cucurbits), *Lagenaria siceraria* (Molina) Standl. (Lauki), *Cucumis sativus* L. (Kheera), *Abelmoschus esculentus* L. (Lady's Finger), *Allium cepa* L. (Onion), *Solanum tuberosum* L. (Potato) and *Raphanus sativus* L. (Radish)]; fodder [*Trifolium alexandrinum* L. (Berseem)]. The natural vegetation [*Dalbergia sissoo* Roxb. (Shisham), *Cassia fistula* L. (Cassia), *Azadirachta indica* A.Juss. (Neem) and *Morus alba* L. (Morus)] and planted plants like *Parkinsonia aculeate* L. (Parkinsonia), *Thevetia peruviana* (Yellow Kaner/Oleander) and *Eucalyptus tereticornis* Sm. (Eucalyptus), *etc.* for greening the area was sparse.

So far, development of the garden for *ex-situ* conservation, plant specimens were collected and brought by altogether 17 Scientists and garden staff at different time period during their services from field, inter-departmental gardens, scientific and educational institutions, Government and reputed nurseries of different parts of the country by various ways including exchange programme, kept in the nurseries to get them adapted to the local environmental conditions. Zonation along periphery of the garden made and planted at varying distance between plants to plants and importance/properties/type-wise specific sections were developed including water bodies and necessary actions taken and duties performed for other activities.

Result and Discussion

To know local environmental conditions of Delhi NCR, meteorological data recorded at Safdarjung Weather Station, New Delhi during last 18 years (1997 to 2014) was collected. The average atmospheric temperature ranged from 2.9°C to 44.7°C, however, during winter season it may be as low as 0.2°C in January with dense fog for 61 days (mostly during January and February) and maximum 48.8°C in May with hot storm. Rainfall is recorded 698 mm for 88 days which occurs mainly during month of July and August. Relative humidity remains 63 per cent whereas wind velocity remains 62 km/h. Thus, weather condition mostly remains dry whereas, dust storms are frequent during summer season. The available water table of BGIR measured by concerned department of U.P. Govt. in between 2006 and 2014 as per which depth ranged from 52.32 ft. to 66.52 ft. during pre-monsoon and from 45.75 ft. to 65.83 ft. during post-monsoon season.

Since the inception of the Garden and as on 31.03.2015 (except the period 2010-13), more than 10,500 (total 10,666) individuals of 876 plant species [comprising Pteridophytes (25), Gymnosperms (11) and Angiosperms (840)] depicted in Table 4.1, among 37 species are listed in Red Data Book (Table 4.2), brought from 23 States of the country are conserved under *ex-situ* Conservation in different components and other activities are as follows:

1. Arboretum (Green Belt, Economic Plants Garden and Fruits Garden)
2. Five Nurseries
3. Thematic Gardens (Cultural Heritage/Historic plants, Map of India, Cactus and Succulent Garden, Medicinal Garden (Ayur Vatika), Landscaped Garden)
4. Horticulture Gardens (Orchids, Ferns, Rosaries, Bougainvillea's and Seasonal Gardens)
5. Water body for Aquatic Plants
6. Other Specific or Specialised Gardens

Arboretum

There are three arboretum in BGIR at present which are as follows:

1. Green Belt

The peripheral areas of this garden along boundary is divided into 8 zones and 92 species of woody plants important among *Dalbergia latifolia* Roxb., *D. sissoo* Roxb., *Hardwickia binata* Roxb. (Arjan or Indian Blackwood), *Hildegardia populifolia*

Table 4.1: Introduction of Plant Species in BGIR since Inception up to March, 2015

Sl.No.	Type of Plants	Total
1	Pteridophytes	25
2	Gymnosperms	11
3	Angiosperms	840
	Aquatic Plants	17
	Bamboo	10
	Cactus	38
	Economic Plant	82
	Fruit	27
	Medicinal	294
	Orchids	36
	Ornamental	149
	Palms	42
	Succulents	53
	Woodland	92
No. of Species		**876**
No. of Individual		**10,666**

The number and types of plants are based on available data on database.

Table 4.2: List of Endemic and Threatened Plant Species Conserved in BGIR since Inception up to March, 2015

Sl.No.	Name of Species	Vernacular Name	Status
1	*Acorus calamus* L., Acoraceae	Bach, Sweet Flag or Calamus	Endemic and Least Concern (IUCN 3.1)
2	*Aerides odorata* Lour., Orchidaceae	Fragrant Aerides	Endemic and Endangered (IUCN 3.1)
3	*Alluaudia procera* (Drake) Drake		New, Near Threatened (IUCN 2.3)
4	*Alstonia macrophylla* Wall., Apocynaceae	Hard alstonia, Hard milkwood	Naturalised and Least Concern (IUCN 2.3)
5	*Asclepias curassavica* L., Apocynaceae	Blood flower	Naturalised and Secure (Nature Serve)
6	*Bambusa vulgaris* Schrad. *ex* J.C.Wendl., Poaceae	Buddha's Belly Bamboo	Endemic, Secure (Nature Serve)
7	*Bambusa vulgaris* var. *vittata* (Rivière and C. Rivière) Mc Clure, Poaceae	Golden Bamboo, Vittata	Naturalised, Secure (Nature Serve)
8	*Couroupita guianensis* Aubl., Lecythidaceae	Cannonball tree, Shivlingi	Naturalised and Least Concern (IUCN 2.3)
9	*Cycas revoluta* Thunb., Cycadaceae		Naturalised and Least Concern (IUCN 3.1)
10	*Cycas sphaerica* Roxb., Cycadaceae	Arjuna chettu, Oruguna	Endemic and Data Deficient (IUCN 3.1)

Sl.No.	Name of Species	Vernacular Name	Status
11	*Dalbergia latifolia* Roxb., Fabaceae	Rosewood, Black rosewood	Endemic and Vulnerable (IUCN 2.3)
12	*Delonix regia* (Boj. ex Hook.) Raf., Fabaceae	Gulmohar	Naturalised and Near Threatened (IUCN 2.3)
13	*Ephedra foliata* Boiss. ex C.A.Mey., Ephedraceae	Shrubby horsetail	Endemic and Least Concern (IUCN 3.1)
14	*Epiphyllum anguliger* (Lem.) G.Don, Cactaceae	Fishbone cactus, Wijayakusuma	Naturalised and Least Concern (IUCN 3.1)
15	*Epiphyllum oxypetalum* (DC.) Haw., Cactaceae	Queen of the night, Brahma Kamalam	Naturalised and Least Concern (IUCN 3.1)
16	*Equisetum hyemale* L., Equisetaceae	Horsetail, Scouring rush	Naturalised and Secure (Nature Serve)
17	*Euphorbia mayurnathanii* Croiz.		Endemic, Extinct in the Wild (IUCN 2.3) and Appendix II of CITES
18	*Euphorbia milii* Des Moul., Euphorbiaceae	Crown of thorns, Christ plant/thorn	Naturalised and Data Deficient (IUCN 3.1)
19	*Euphorbia milii* var. *longifolia* Rauh, Euphorbiaceae	Crown of thorns, Christ plant/thorn	Naturalised and Data Deficient (IUCN 3.1)
20	*Euphorbia tirucalli* L., Euphorbiaceae	Satala, Pencil Cactus, Milk Bush	Naturalised and Least Concern (IUCN 3.1)
21	*Ginkgo biloba* L., Ginkgoaceae	Maidenhair tree	Naturalised and Endangered (IUCN 2.3)
22	*Hamelia patens* Jacq., Rubiaceae	Fire bush, Humming bird Bush	Naturalised and Not evaluated (IUCN 3.1)
23	*Heritiera littoralis* Dryand., Aiton (Malvaceae)		New Least Concern (IUCN 3.1)
24	*Hildegardia populifolia* (Roxb.) Schott. et Endl., Malvaceae	Gali Buddaga, Poplar leaved Ardor	Endemic and Critically Endangered (IUCN 2.3)
25	*Hyophorbe lagenicaulis* (L.H.Bailey) H.E.Moore, Arecaceae	Bottle Palm	Naturalised and Critically Endangered (IUCN 2.3)
26	*Hypericum gaitii* Haines, Hypericaceae	Gaitii	Endemic and Endangered
27	*Lasiococca comberi* Haines, Euphorbiaceae		Endemic and Intermediate
28	*Madhuca nerifolia* (Moon) H.J.Lam, Sapotaceae	Mahua, Illipe Butter Tree	Naturalised and Endangered (IUCN 2.3)
29	*Pittosporum eriocarpum* Royle Pittosporaceae		Endemic (Uttarakhand and UP) and critically endangered (IUCN 2.3)
30	*Pteris ensiformis* Burm. f., Pteridaceae	Silver lace fern, Sword brake fern	Endemic and Secure (Nature Serve)
31	*Pterocarpus marsupium* Roxb., Fabaceae	Vijayasar or Indian Kino Tree	Endemic and Vulnerable (IUCN 2.3)
32	*Pterocarpus santalinus* L.f., Fabaceae	Red Sanders, Red Sandal wood	Endemic and Endangered (IUCN 2.3)

Sl.No.	Name of Species	Vernacular Name	Status
33	*Santalum album* L. Santalaceae		Endemic and Vulnerable (IUCN 2.3)
34	*Saraca asoca* (Roxb.) Wilde, Fabaceae	Ashoka tree	Endemic and Vulnerable (IUCN 2.3)
35	*Swietenia macrophylla* King, Meliaceae	Mahogany, Big-leaf Mahogany	Naturalised and Vulnerable (IUCN 2.3)
36	*Swietenia mahagoni* (L.) Jacq., Meliaceae	West Indies Mahogany	Naturalised and Endangered (IUCN 2.3)
37	*Washingtonia filifera* (Lindl.) H.Wendl., Arecaceae	California fan palm and Petticoat palm	Naturalised and Near Threatened (IUCN 2.3)

(Roxb.) Schott. and Endl., *Lannea coromandelica* (Houtt.) Merr., *Mimusops elengi* L. (Molsri), *Pterocarpus marsupium* Roxb., *Pterocarpus santalinus* L.f., *Schleichera oleosa* (Lour.) Merr. (Kusum), *Shorea robusta* Roth, Tectona grandis, *Terminalia arjuna* (Roxb.) Wight and Arn., *Wrightia tinctoria* (Roxb.) R.Br. and other species have been planted to create thick green belt all around to mitigate the ambient air pollutants and to prevent entry of hot air storm during summers and cold wave during extreme cold.

2. Economic Plants Garden

Eighty two species of plants having economic importance as sources of food, medicine, fuel, oil, timber, fibres, *etc.* have been planted in this section. Important among them are *Albizia amara* (Roxb.) Boiv., *Albizia lebbeck* (L.) Benth., *Albizia odoratissima* (L.f.) Benth., *Albizia procera* (Roxb.) Willd., *Bombax ceiba* L., *Bombax insigne* Wall., *Dalbergia latifolia* Roxb., *Dalbergia sissoo* Roxb., *Dipterocarpus grifthii* Miq., *Dipterocarpus indicus* Beddome, *Ficus roxburghii* Wall., *Garcinia indica* (Thouars) Choisy, *Gmelina arborea* Roxb., *Grevillea robusta* A.Cunn. ex R.Br., *Holigarna arnottiana* Wall. ex Hook. f., *Jatropha glandulifera* Roxb., *Jatropha gossypifolia* L., *Jatropha integerrima* Jacq., *Litsea glutinosa* (Lour.) C.B.Rob., *Mullotus phillppensis* (Lam.) Muell. Arg., *Oroxylum indicum* (L.) Benth. ex Kurz, *Pongamia pinnata*, *Prosopis cineraria* (L.) Druce, *Pterocarpus santalinus* L.f., *Rubia cordifolia* L., *Salix alba* L., *Shorea robusta* Roth, *Shorea roxburghii* G. Don, *Sterculia alata*, *Sterculia urens* Roxb., *Swietenia macrophylla* King, *Swietenia mahagoni* (L.) Jacq., *Tectona grandis* L.f., *Terminalia alata* Heyne ex Roth, *Terminalia arjuna* (Roxb.) Wight and Arn., *Terminalia bellirica* Gaertn.) Roxb., *Terminalia catappa* L., *Terminalia chebula* Retz., *Terminalia tomentosa* L., *Thespesia populnea* (L.) Sol. ex Corrêa and other species.

3. Fruits Garden

Twenty seven species of fruit yielding plant have been planted. Important among them are *Annona reticulata* L., *Annona squamosal* L., *Anthocephalus cadamba* (Roxb.) Miq., species of *Citrus* like Chakotra, Malta, Mosambi, lemon, Kinnow, *Ficus glomerata* Roxb., *Grewia calophylla* Kurz ex Mast., *Litchi chinensis* var. 'Allahabad' and 'Dehradun', *Madhuca indica* J.F. Gmel., *Mangifera indica* L. var. 'Almada' and 'Baganpali', *Musa velutina* H.Wendl. and Drude, Naspathi (China), *Phyllanthus emblica* L., *Prunus cerasoides* D.Don (White var.), *Prunus dulcis* (Mill.) D.A.Webb Badam, *Psidium guajava* L., *Punica granatum* L., *Pyrus pashia* L., *Quercus sapota*, *Syzygium cumini* (L.) Skeels., *Tamarindus indica* L., *Trewia nudiflora* L.and others.

Nurseries

Presently, there are 5 nurseries in this garden, established for varying purposes depending upon plant types and habits of plants. Nursery No. 1 is provided with sprinkler and fogging system for Ferns and Orchids which require little moist climatic condition. Nursery No. 2 is used for nursing of plants known to occur during different Era *i.e. Psilotum nudum* (L.) P. Beauv. and *Equisetum hyemale* L. of Silurian and Devonian and Carboniferous period respectively of Paleozoic Era and *Gingko biloba* L. of Jurassic period of Mesozoic Era as well as other Cycads like *Cycas beddomei* Dyer, *Cycas revoluta* Thunb., *Cycas rumphii* Miq., *Cycas circinalis* L., *Zamia furfuracea* L.f. and other plants having medicinal properties including aquatic plants like *Nymphaea caerulea* Sav. (Blue)], *Thalia geniculata* L. (Fire flag), *Echinodorus palifolius* (Nees and Mart.) J.F.Macbr. (Mexican sword plant), *Hydrilla verticillata* (L.f.) Royle. Nursery No. 3 is for all kinds of plants which require less sun exposure widely called as shade loving plants and others whereas, Nursery No. 4 is specifically dedicated for multiplication and propagation of plants either from seeds, cutting, bulbs, tubers, rhizomes, *etc.* required for conservation, landscaping, distribution during Van Mahotsav and other purposes and Nursery No. 5 is being used for research and development work as well as getting introduced plant species adapted to local environmental condition of this garden.

Thematic Gardens

1. Cultural Heritage/Historic Plants

Other than economic importance, most plant species are connected with religion, cultural events and celebrations, as they are also considered main foundation stone of all civilization. Hence, initially some of rich heritage plants having historic significance as well as known to occur during different Era *i.e. Psilotum nudum* (L.) P. Beauv. and *Equisetum hyemale* L. of Silurian and Devonian and Carboniferous period respectively of Paleozoic Era and *Gingko biloba* L. of Jurassic period of Mesozoic Era. *Adansonia digitata* L. widely called as 'Kalpavriksha' known to appear during Samudra Manthan and various other plants species like *Ficus benghalensis* L. (Banyan), *Ficus religiosa* L. (Pipal), *Mangifera indica* L. (Mango), *Azadirachta indica* A.Juss. (Neem), *Musa paradisiaca* L. (Banan), *Saraca asoca* (Roxb.) Wilde (Ashok), *Santalum album* L. (Chandan), *Ficus benghalensis* var. *krishnae* (C. DC.) Corner (*Krishna Cup*), *Nyctanthes arbor-tristis* (*Parijat* or Harsingar), *Prosopis cineraria* (L.) Druce (Shami), *Nymphaea caerulea* Sav. (Blue) *etc.* having medicinal properties described in Charak Samhita, Epics of Ramayan and Mahabharat, *etc.* have also been conserved and displayed at different locations of the garden.

2. Map of India

This section was developed during 2013 in central part of landscaped area of BGIR to display whole objective of this garden at one place by considering three things:

(*a*) Conservation of endemic and threatened plants of the country.

(*b*) Awareness about plants and flowers declared by different States and Union Territories of India (*c*) Beautification and recreation for visitors along with awareness.

The 'Map of India' (size 20.4 m in length and 18.4 m in width) was conceived in six month referring 'Map of India' released by Survey of India and finalised after discarding different developed designs keeping in view of combination of colour of plants, their heights, foliage, rate of defoliation, mode of maintenance, *etc.* and drawn by own. The territorial area of each State and Union Territory is filled by 21 different plant species including medicinal plants like *Alternanthera versicolor* R.Br. (Lal Badshah), *Asparagus setaceus* (Kunth) Jessop (Satmul), *Brassica juncea* (Ornamental Mustard 'Red Giant'), *Calliandra haematocephala* Hassk., *Carpobrotus edulis* (L.) N.E. Br (Ice Plant), *Chlorophytum brivilianum* (White Musli), *Clerodendrum inerme* (L.) Gaertn. (inerme), *Duranta erecta* L. Varigated, *Epipremnum aureum* (L.) Engl. (Money plant: Green), *Euphorbia cotinifolia* L. (Euphorbia: Red), *Ficus panda* (Panda), *Hamelia patens* Jacq. (Hamelia or Firebush), *Jatropha integerrima* Jacq. (Jatropha), *Pithecellobium dulce* (Roxb.) Benth. (Jalebi), *Putranjiva roxburghii* Wall. (Garvkar), *Sansevieria trifasciata* Prain (Sanseviera), *Sphagneticola trilobata* (L.) Pruski (Wedelia trilobata), *Thevetia peruviana* (Pers.) K. Schum. (Yellow Oleander or Lucky Nut or Kaner), *Thuja occidentalis* L. (Thuja), *Tradescantia pallida* (Rose) D.R.Hunt (Lal Patti) and *Vernonia elaeagnifolia* DC. (Parda bel or Curtain Creeper) developed in the garden itself. During winter season, outer periphery is ornamented either by *Mesembryanthemum criniflorum* L.f. (Ice Plant) and or by *Lobularia maritime* (L.) Desv. (Allysum) and other plants. So far, any botanic garden has yet not made such a huge map of their country using live plants. The territorial boundary has also been illuminated by installing LED light and switch on during night for showing a glimpse of India in tri-colour for commuters of Metro and Aeroplane. Besides, three side of the Map, ornamental gabions have been made for plantation of declared plants and flower of respective State of India for conservation, beautification and awareness purposes.

3. Cactus and Succulents Garden

Plants which are known to grow usually in xeric and other unusual conditions widely called as Cactus and Succulent have also been conserved in Cactus House and Succulent Garden which was made during 2013 in an area measuring 75 m × 70 m (length and width) and has been fenced from three side by 2.40 m wide ornamental fence using 13 different species of Succulents and Cactus and decorated with potted and seasonal plants time-to-time.

The Cactus House contains all together 36 species belonging to four different kinds of Cactus *i.e.* round, column, creeper and aerial including ancestor of Cactus *i.e. Pereskia grandifolia* Haw. (Rose Cactus). Important conserved Cactus species are *Echinopsis calochlora* K. Schum., *Epiphyllum anguliger* (Lem.) G.Don, *E. oxypetalum* (DC.) Haworth, *Ferocactus latispinus* Britton and Rose, *Gymnocalycium mihanovichii* (Fric ex Gürke) Britton and Rose, *Lophophora williamsii* (Lem.) J.M.Coult., *Mammillaria baumii* (Boedeker), *M. phitauiana* (Embaxter) Werderm. Ex Backeb., *Opuntia ficus-indica* (L.) Mill., *O. littoralis* (Engelm.) Cockerell, *O. microdasys* (Lehm.) Pfeiff. and others.

Outsides Cactus House and within premises more than 50 species of succulents important among them *Agave* sp. Variegated, *A. angustifolia* Haw., *A. salmiana* Otto ex Salm-Dyck, *A. vivipara* L., *Aloe vera* (L.) Burm. (White), *A. vera* var. *chinensis* (Dark green), Beaucarnea recurvate Lem., *Bryophyllum pinnatum* (Lam.) Kurz., *B. uniflorum* (Stapf) A. Berger (Kalanchoe), *Cereus peruvianus* (L.) Mill., *Dracaena braunii* Engl. (Lucky bamboo), D. *fragrans* (L.) Ker Gawl., *D. marginata* Lam., *D. marginata* var. *bicolor* Lam., *D. marginata* var. *colorama* Lam., *D. marginata* var. *tarzan* Lam., *D. reflexa* Lam., *Euphorbia abschneiden, E. franckiana* A.Berger, *E. lacteal* Haw., E. *leuconeura Boiss., E. milii* Des Moul., *E. tirucalli* L., E. tithymaloides L. (Pedilanthus), E. trigona Mill., *Furcraea foetida* (L.) Haw., *F. foetida* (L.) Haw. 'Mediopicta', *F. foetida* (L.) Haw. Variegated, *Huernia zebrine* N.E.Br., *Jatropha curcas* L. (Jatropha) Green, *Kalanchoe blossfeldiana* Poelln (Kalanchoe), *Pedilanthus bracteatus* (Jacq.) Boiss., *Portulaca grandiflora* Hook., *P. afra* Jacq. (Jade plant), *Ravenala madagascariensis* Sonn., *Sansevieria cylindrical* Bojer, *S. trifasciata* Prain, *S. trifasciata* var. *hahnii* Dwarf plant, *S. zeylanica* (L.) Willd. (Without Yellow Strip at Edge), *Stapelia gigantea* N.E.Br., *Yucca aloifolia* L., *Y. capensis* L.W.Lenz, *Y. gigantean* Lem. 'Jewel and other species have been planted.

4. Medicinal Garden (*Ayur Vatika*)

Most plant species possess certain potent chemical constituents in the form of secondary metabolites *i.e.* organic compounds including terpenoids, special nitrogen metabolite (like non-protein amino acids, amines, cyanogenic glycosides, glucosinolates and alkaloids) tannins, resins, taraxerol, phenolics, volatiles oils and others which individually and or in combination if used/applied cure illness/disease and thus called as medicinal plants. So far, 294 such medicinal plants have been introduced in this garden. Broadly, in designated '*Ayur Vatika*', plants which have potent properties to treat illness/diseases of human have been categorised into 8 different categories *i.e.* Digestive System, Blood and Circulatory System, Musculo-Skeletal System, Skin Diseases, Urino-Genital System, Respiratory System, Fever and Inflammation and Nervous System and accommodated more than 100 species and are placed in different locations. *Abrus precatorius* L. (Ratti), *Aloe vera* (L.) Burm. (White) (Aloe), *Bacopa monnieri* (L.) Pennell (Brahmi), *Barleria prionitis* L. (Vajradanti), *Cascabela thevetia* (L.) Lippold (Kaner Yellow), *Chlorophytum borivilianum* L. (Safed Musali), *Cinnamomum camphora* (L.) J.Presl. (Camphore), *Cissus quadrangularis* L. (Harjod), *Clitoria ternatea* L. (Aparajita), *Curcuma longa* L. (Haldi), *Cymbopogon schoenanthus* (L.) Spreng. (Lemon Grass), *Drypetes roxburghii* (Wall.) Hurus. (Putranjiva), *Elaeocarpus ganitrus* Roxb. ex G.Don (Rudraksh), *Hibiscus rosa-sinensis* L. (Hibiscus), *Moringa oleifera* Lam. (Sahjan), *Nerium oleander* L. (Kaner Lal), *Plantago major* L. (Isabgol), *Psoralea corylifolia* L. (Babchi), *Pterocarpus santalinus* L.f. (Chandan Lal), *Rauvolfia serpentine* (L.) Benth. *ex* Kurz (Rauvolfia), *Santalum album* L. (Chandan Safed), *Saraca asoca* (Roxb.) Wilde (Ashok), *Tinospora cordifolia* (Thunb.) Miers (Giloy), *Vitex negundo* L. (Nirgundi) and other well established plant species which are enlisted in Ayurveda, Unani, Siddha and other treatment systems are planted in this garden.

5. Landscape Garden

As BGIR is a botanic garden, artistic landscape garden with topiary is usually not required. However, at two different locations of the garden, landscape garden has been developed with lush green lawn and planting plants of landscaping importance like *Cycas revoluta* Thunb. (Cycas), *Furcraea foetida* (L.) Haw. var. *mediopicta* (Furcraea), *Hyophorbe lagenicaulis* (L.H.Bailey) H.E.Moore (Bottle palm), *Washingtonia robusta* H.Wendl. (Washintonia) and *Yucca capensis* Lenz (Yucca) with or without mounds and hedge of *Alternanthera versicolor* R.Br. (Lal Badshah), *Clerodendrum inerme* (L.) Gaertn. (Inerme), Duranta erecta L. (Duranta), *Ficus benjamina* L. (Benjamina), *Ficus panda, and Hamelia patens* Jacq. (Hamelia), *etc.* to enhance aesthetic view and to meet psychological requirements/reduce stress of visitors during their stay.

Horticulure Gardens

1. Orchidarium

Altogether 36 Orchids have been conserved in Nursery No. 1, among important species are *Cateleya* sp., Coelogyne flaccida Lindl., *C. stricta* (D. Don) Schltr., *Cymbidium barbatulum* Lindl., *Dendrobium densiflorum* Lindl., *D. nobile* Lindl., *Eria vitata* Lindl., *Vanda poepoe* var. *diana, Vanda tessellata* (Roxb.) Hook. ex G.Don, and others.

2. Ferns

Altogether 25 species of ferns have been conserved in Nursery No. 1, among *Psilotum nudum* (L.) P. Beauv. and *Equisetum hyemale* L. of Silurian and Devonian and Carboniferous period respectively of Paleozoic Era, species of *Pteris, Adiantum,* Sword of Tipu Sultan and others have been planted.

3. Rosaries

During 2014, this section was developed by planting more than 250 individuals of wild rose plant including a few ornamental of 15 colours. Important among them *Rose canina* L., *R. chinensis* Jacq., *R. chinensis* var. *viridiflora, R. clinophylla* Thory, *R. gallica* L., *R. roxburghii, Rosa alba, Rosa arvensis* Hudson and other species were planted.

4. Bougainvilleas

During 2014, a Bougainvillea Section was also developed by planting more than 150 stalks of different colour Bougainville and was further strengthened during Van Mahotsawa'2015.

5. Seasonal Flower Gardens

Other than perennial ornamental species, more than 35 species winter annuals and 14 species of summer annuals are planted annually in respective seasons in beds of different sections, along road and pots for display and to enhance aesthetic view of the garden.

Water Body for Aquatic Plants

Presently, there are two water bodies existing in this garden and accommodating altogether 17 species of aquatic plants namely *Nelumbo nucifera* Gaertn. (Lotus or Kamal), 5 species of *Nymphaea* (Kumud) [(namely *Nymphaea pubescens* Willd.

(White), *N. rubra* Roxb. ex Andrews (Red), *N. mexicana* Zucc. (Yellow), *N. colorata* Peter (Voilet/Purple) and *N. caerulea* Sav. (Blue)], *Thalia geniculata* L. (Fire flag), *Echinodorus palifolius* (Nees and Mart.) J.F.Macbr. (Mexican sword plant), *Hydrilla verticillata* (L.f.) Royle, *Pistia stratiotes* L., *Salvinia minima* Baker, *Vallisnaria* sp., *Potamogeton crispus* L., Hinche, and *Chara* (Macro Algae).

Other Specific Components or Specialised Gardens

To create specific component or specialised gardens of different kinds of plant like Cycads including other Gymnosperms, bamboos, palms and pandanus (screw palm), Cycadarum, Bambusetum, Palmetum and Pandarum respectively as well as other sections like *Bonsai,* Fragrance and Beverage gardens, respective plant species have been conserved in nurseries and once the landscape of whole garden is prepared they will be planted in respective places systematically. For above purposes, the specific plant group are as follows:

Cycad Collection

Presently, it has been created in Nuresry No. 2 and containing altogether 11 species among Cycads like *Cycas beddomei* Dyer, *Cycas revoluta* Thunb., *Cycas rumphii* Miq., *Cycas circinalis* L., *Zamia furfuracea* L.f. are important as they are listed in RED Databook whereas *Gingko biloba* L., a 'living fossil' known to occur during Jurassic period of Mesozoic Era. Besides, species of *Thuja occidentalis* L., *Araucaria* and *Juniperus chinensis* L. are also planted at different locations for landscaping and beautification.

Bambusetum

Altogether 10 species have been introduced, among Budha bamboo, golden bamboo and others are important.

Palmetum

Altogether 42 species were introduced, among *Hyophorbe lagenicaulis* (L.H.Bailey) H.E.Moore (bottle palm), *Roystonea regia* (Kunth) O.F.Cook, *Washingtonia robusta* H.Wendl., *Phoenix dactylifera* L., *Phoenix roebelenii* O'Brien, *Phoenix sylvestris* (L.) Roxb. and others are important.

Bonsai

As this garden is meant for functioning as a conservatory of live endemic and threatened plants of the country in their own form however, on receipt of two bonsai of *Ficus benghalensis* (Banyan or Bar) and *Carissa carandas* L. (Karonda) which are nearly 18 and 12 year old respectively were donated by Dr. Binoy Prasad, Howrah on 07.09.2015.To meet the requirement of visitors and for creating general awareness, bonsai specimens were displayed.

Other Functional Components

Herbarium

Under scientific research programme on 'Flora of Delhi', projects and dissertation work of students and scholars so far 1,800 Herbarium including specimens of this garden prepared, studied and preserved for reference, training and awareness purposes.

Seed Bank

The Seed Bank of this garden was established during 2006 with a view to conserved seeds of endemic and threatened plant species including other important species, periodic germination testing of post-storage seeds of different time period and survival of their seedling(s) in the field. So far, seeds of 257 species comprising have been conserved.

Database

The work for preparation of Database of introduced plants started during 2006 and so far, more than 60 per cent work for 15 fields about specific species with photograph has been completed for Medicinal including other introduced plant species.

Soil Testing Laboratory

A soil testing laboratory was inaugurated by Dr. (Mrs.) R. Dalwani, Advisor, MOEF and CC on 17.07.2015, where one can analyse at least 15 physico-chemical characteristics of soil to know its suitability/status for plantation with or without treatment.

Rescue of Plants/Trees by

i. Transplantation of Trees

97.5 per cent of 160 plant individuals of girth size >10 cm dia. (among total 1,235 plants) having ornamental, economic importance, endemic as well as endangered/critically endangered (as per IUCN) in nature were successfully transplanted at different locations in the garden from alignment of new Metro Route of Phase – II. All such transplantation was carried out under supervision of Dr. Sheo Kumar, Sci. 'E' during Feb. and Mar., 2015 is one of the greatest achievements of whole infrastructural developmental project so far reported.

ii. Conservation of Plants/Trees Received on Donation, if any, found important

Most of nature and plant lovers grow plants in their houses, farmlands, terrace and other places as per their wish and interest by procuring them from various places/sources, nurture them like kids but unable to maintain after certain time period of their life. Thus, they remain unmaintained, die pre-maturely and at times, space/area occupied by them, vacated for some other purposes by throwing them brutally. BGIR always welcome such plants and conserved them if, found uncommon, rare, endemic and as per requirement of the garden.

Training and Environmental Awareness Programmes

Educational training with respect to conservation of plants, transplantation of trees, scientific research in the form of 'projects' and 'dissertation work' of students and scholars, identification of specimens, authentication of plants is being carried out along with necessary practical and demonstration as and when required. However, on permanent basis for training and awareness purposes and to replace use of chemical fertilizers (usually applied for growth of plants), prototype Vermi and Organic Composting units were established separately.

Vermi and Organic Composting

By applying techniques and process of Vermi and Organic Composting, the generated garbage, leaf litter, *etc.* of this garden is converted into organic composts on regular basis and used in the field for developmental work. Under awareness programme of the garden, visitors are keeping aware through installed signage depicting process for preparation of composts, machine, tools/appliances and specific pits made for both units.

Besides, being a National level garden, contributions also made time-and-again during various decision making meetings held for development of the country as well as conservation of plants. Annually, different National and International events like National Science Day and International Day for Biological Diversity, *etc.* are being organised on respective dates for following activities:

- *Van Mahotsav* by Sapling Distribution and Plantation Programme,
- Orchid and Flower Shows,
- Film Shows, Nature Walk,
- Science Quiz, Drawing and Debate Competitions, *etc.*

BGIR fulfilling following Articles of CBD, 1992 after its Establishment

The Convention of Biological Diversity (CBD) was signed during Earth Summit in 1992 and recommendations came out in the form of 42 Articles, enforced globally in 1993 for implementation in signatory countries. As BGIR was established after above convention, hence, so far fulfils 10 out of 42 Articles (BGIR, 2015) are as follows:

- Article 6: National Strategies on Conservation of Biodiversity and Sustainable Use.
- Article 7: Identification and Monitoring.
- Article 8: *in-situ* Conservation.
- Article 9: *ex-situ* Conservation.
- Article 10: Sustainable Use of Components of Biological Diversity.
- Article 12: Research.
- Article 13: Training, Public Education and Awareness.
- Article 15: Access to Genetic Resources (and benefit sharing).
- Article 17: Exchange of Information.
- Article 18: Technical and Scientific Co-operation.

Future Development

The following components are proposed for creation along with strengthening of the existing:

1. State of Art Infrastructure like

i. Administrative and Research Buildings including other infrastructures for *ex-situ* Conservation and R&D work:

- ✰ Laboratories for Environmental and Biological Studies including Plant Pathology, Biotechnological studies with Tissue Culture and Integrated Pest Control Management with Incinerator
- ✰ Herbarium
- ✰ Storage facilities for Conservation of Seeds, Pollen and Tissues
- ✰ Library
- ✰ Interpretation Centre
- ✰ Conservatories as per Phyto-geographical Areas of the Country.
- ✰ Amphitheatre, Conference/Seminar Halls, Museum and Exhibition Halls
- ✰ Kiosk, Sale Counters, Cafeteria, Public Utility Centres and Rest Room

ii. Other basic components for Conservation, Development and Maintenance:

- ✰ Drip and Sprinkler Irrigation System
- ✰ Operation of Garden Appliance, Eco-Friendly Vehicle, *etc.*
- ✰ Power Back-up and Solar Lights
- ✰ Security Surveillance System with CCTV Camera, Audio-Video and Biometry
- ✰ Drinking water supply and Sanitation
- ✰ Water bodies and Fountains
- ✰ Generation of Vermi and Organic Composts, *etc.*

2. Training and Capacity Building

3. Eco-Tourism and Recreation

Acknowledgement

Author is thankful to the Director, Botanical Survey of India, Kolkata for constant inspiration and encouragements and to the Ministry of Environment, Forest and Climate Change for providing logistic support and opportunity to look after day-to-day development and maintenance work of Nation's pride.

References

BGIR 2015. Brochure on Memorandum of Understanding between MOEF and CC and NOIDA. June 5th, 2015. Botanic Garden of Indian Republic, Govt. of India, NOIDA.

BSI 2015. Plant Discoveries: New Genera, Species and New Records. Botanical Survey of India, MOEFC and CC, Govt. of India, Kolkata.

CI 2007. "Hotspots by Region, Biodiversity Hotspots (Conservation International)". Online.

Clémentin-Ojha, C. 2014. 'India, that is Bharat…': One Country, Two Names. *South Asia Multidisciplinary Academic Journal,* Online.

Crame J.A. and Owen A.W. 2002. Palaeobiogeography and Biodiversity Change: The Ordovician and Mesozoic–Cenozoic Radiations. Geological Society of London, Geological Society Special Publication.

Heitzman J. and Worden R.L. 1996. India: A Country Study. Area Handbook Series, Washington, D.C.

Karanth K.P. 2006. Out-of-India Gondwanan Origin of Some Tropical Asian Biota. *Current Science* 90(6): 789–792.

McWhirter N. and McWhirter R. 1964. Dunlop illustrated encyclopedia of facts. Dreghorn Publications Ltd., New York.

Roach J. 2005. Conservationists Name Nine New: Biodiversity Hotspots. National Geographic News.

Tritsch M.F. 2001. Wildlife of India. Harper Collins, London.

Chapter 5

Contributions of National Botanic Garden, Trombay (NBGT) in *In-situ* vis-à-vis *Ex-situ* Conservation

H.A. Barbhuiya and C.K. Salunkhe

Landscape and Cosmetic Maintenance Section, Architectural and Structural Engineering, Division, Bhabha Atomic Research Centre, Trombay, Mumbai – 400 085, Maharashtra
****E-mail: hahmed@barc.gov.in, chandrak@barc.gov.in***

Abstract

An attempt has been made to appraise the contributions of NBGT towards plant conservation. In the current census we have found ca. 1500 species of vascular plants belonging to 130 families and 562 genera are occurring within the ambit of NBGT. Out of which, the garden is supporting in-situ conservation of ca. 600 species of native plants by preserving their original natural habitats, of which ca. 70 species are endemic to Western Ghats and Peninsular India. Besides, the garden is also supporting ex-situ conservation of ca. 900 species of vascular plants, which includes many native as well as rare exotic species. Interestingly, in the course of time a considerable percentage of them are well established and some of them are being naturalized.

Keywords: Plant conservation, Introduction, Naturalization, Threatened, Endemic.

INTRODUCTION

It is a well-established fact that worldwide many thousands of plant species are currently threatened with extinction (Gomez-Mejia *et al.*, 2006). In a recent study it was found that the current extinction rates are 1,000 times higher than natural background rates of extinction and future rates are likely to be 10,000 times higher (de Vos *et al.,* 2014). Therefore, the future survival of humanity will only depend on the conservation and protection of natural wealth, and the extinction of a species or a genetic line symbolizes the loss of a unique treasure, which has evolved as a result of a long evolutionary process (Bapat *et al.,* 2012). Botanic gardens are long been involved in the introduction and dissemination of plant resources. Conservation of plant diversity is already, and very appropriately, recognized as being a major activity for botanic gardens through their research and educational programs (Ashton, 1988). Currently there are 1775 botanic gardens and arboreta in 148 countries around the world with many more under construction or being planned (BGCI, 2016).

According to a recent estimates there are *ca.* 19510 spp. (1274 sp. of Pteridophytes, 77 sp. of Gymnosperms and 18159 sp. of Angiosperms) of vascular plants occurring in India (Singh and Dash, 2015) out of which 4381 taxa (*ca.* 22.4 per cent) are endemic to India (Singh *et al.,* 2015). However, due to continuous habitat fragmentation by indiscriminate felling of forest trees, clearance of forest cover for expansion of agricultural land, encroachment of forest land, illegal trade for medicinal use, mining and various other anthropogenic activities, many vascular plants of the country are become much rarer, highly threatened and in some cases already been extinct. Nayar and Sastry, (1987-1990) in *"The Red Data Book of Indian Plants"* enlisted *ca.* 622 species of threatened vascular plant. Later, in the year 2003 it was found that the total number threatened vascular plants rose to 1236 (Rao *et al.,* 2003), but in the recent estimate this red figure climbed up to 2121 (Arisdason and Lakshminarasimhan, 2016), *i.e.* approximately 10.8 per cent of the Indian vascular plant species are now became threatened and their numbers are constantly increasing day by day. In India there are *ca.* 122 Botanic Gardens and it is presumed that these botanic garden are supporting living collections of 7,000 to 10,000 spp. of vascular plants (CBD, 2016), which includes many exotic species. But the actual number of indigenous plants conserved in these botanic gardens is expected to be very low as only few botanic gardens are currently focusing on conservation issues of threatened and endemic plants of the country.

To safeguard these threatened and endemic plants of the country and especially of Western Ghats as one of its prime objective *"The National Botanic Garden, Trombay"* (NBGT) was came into existence at the campus of Bhabha Atomic Research Centre in 1958. The garden is spread over 8.7 sq. km area out of which *ca.* 77 per cent area is under natural forest cover comprising Mandala and Trombay Hills. Besides above, the garden has a large stretch of mangrove vegetation comprising an additional area of *ca.* 4.04 sq. km. This virgin forest cover is totally secluded from urbanization and in a way it is a 'bio-reserve'. On the other side, since inception of NBGT, many botanists and horticulturalists were involved in introduction of native as well as cultivated ornamental plants to this garden and in the course of time many of them are being naturalized.

The natural vegetation's of NBGT can be classified into following three categories 1. Tropical dry deciduous forests, 2. Tropical semi-evergreen forests and 3. Littoral and swamp forests. The vegetation's of the hills are mainly dominated by tropical dry deciduous forests and the major key components of this kind of forests are: *Bombax ceiba, Bridelia retusa, Dalbergia latifolia, Ficus arnottiana, Firmiana colorata, Garuga pinnata, Gliricidia sepium, Grewia tiliifolia, Holoptelea integrifolia, Lagerstroemia parviflora, Lannea coromandelica, Pterocarpus marsupium, Schleichera oleosa, Sterculia urens, Stereospermum tetragonum, Tectona grandis* and *Wrightia tinctoria*. Whereas, tropical semi-evergreen type of forests are occasionally found in patches at the foot hills. The major components of tropical semi-evergreen forests are: *Azadirachta indica, Bauhinia racemosa, Borassus flabellifer, Caryota urens, Ficus hispida, F. racemosa, Grewia serrulata, Ixora brachiata, I. pavetta, Mallotus philippensis, Manilkara hexandra, Muntingia calabura, Pongamia pinnata, Putranjiva roxburghii, Sapindus laurifolius, Syzygium cumini* and *Trema orientalis*. While, the littoral and swamp forests are found along the sea coasts, the dominant taxa of this kind of forest are: *Aegiceras corniculatum, Avicennia officinalis, A. marina, Lumnitzera racemosa, Rhizophora mucronata* and *Sonneratia apetala*. The climate of NBGT is tropical and it experiences four seasons during a year. March to May is the summer season followed by rainy season from June to September. During this period rainfall varies from 144 to 269 cm. The post monsoon season is October to November and December to February is the winter. The temperature of the area varies from 15°C to 40°C, the hottest months are April and May. The altitude varies from sea level to up to 300 m above MSL and relative humidity varies from 67 per cent to 86 per cent.

Methodology

Frequent field visits were conducted at regular intervals from April 2014 – May 2016 to explore all areas of NBGT *viz*. Trombay hill, Mandala Hill, BARC Premises and Anushakthinagar colony for collection of plant specimens, photography and recording of morphological as well as phenological data. All the collected plant specimens were preserved in the form of herbarium sheets by following Jain and Rao (1977). The identity of collected plant specimens were determined in consultation with authentic taxonomic literatures (*e.g.* Hooker, 1872-1897; Cooke, 1901-1908; Almeida, 1996-2010; Sharma *et al.*, 1996; Abraham, 1997 and 1998; Singh and Karthikeyan, 2000; Singh *et al.*, 2001; Mishra and Singh, 2001; Ogale, 2010 *etc.*). Finally their identity was verified with authentic herbarium specimens housed at HBARC (Herbarium of Bhabha Atomic Research Centre, Mumbai). All the studied voucher specimens were also deposited at HBARC.

Results and Discussion

The present study had revealed the occurrence of *ca*. 1500 species belonging to 130 families and 562 genera, of which 600 species are naturally occurring in wild and the rest 900 species are being introduced (Table 5.1). Among wild flora the family Fabaceae is represented with highest number of taxa *i.e*. 90 species belonging to 40 genera, followed by Poaceae with 50 species belonging to 35 genera, followed by Asteraceae with 28 species and 19 genera, Euphorbiaceae with 27 species and 17 genera, Malvaceae with 27 species and 15 genera, Acanthaceae with 23 species and 16 genera, Convolvulaceae with 22 species and 10 genera, Rubiaceae with 21 species and 14 genera, Cyperaceae with 21 species and 5 genera and so on (Table 5.2).

Table 5.1: Present Status of the Vascular Olants occurring at NBGT

Taxon	Wild	Introduced	**Total**
Family	90	100	130
Genera	345	465	562
Species	600	900	1500

Table 5.2: Twelve Dominant Families of Wild Flowering Olants occurring at NBGT

Sl.No.	*Family*	*Species*	*Genera*
1	Fabaceae	90	40
2	Poaceae	50	35
3	Asteraceae	28	19
4	Euphorbiaceae	27	17
5	Malvaceae	27	15
6	Acanthaceae	23	16
7	Convolvulaceae	22	10
8	Rubiaceae	21	14
9	Cyperaceae	21	5
10	Apocynaceae	13	11
11	Amaranthaceae	11	7
12	Commelinaceae	10	3

Contribution towards *in-situ* Conservation

In the present census we have found that NBGT is supporting *in-situ* conservation of *ca.* 600 species of native plants belonging to 345 genera and 90 families in the natural virgin forests of Mandala and Trombay Hill, out of which *ca.* 12 per cent (*i.e.* 70 taxa) are endemic to Western Ghats and Peninsular India (Figure 5.1, Table 5.3). Among them, *Firmiana subglabra* (V.Abraham and Dutt) Kosterm. is strictly endemic to Trombay Hill of NBGT. This species is currently known from a single location and its area of occupancy estimated to be less than 1 km^2 and population size is estimated to lower than 10 mature individuals. Hence, following IUCN Red List Categories and Criteria (IUCN, 2001) this taxon is classified as Critically Endangered (CR B2a+D). During the study it was found that following 13 endemic taxa are rare in their natural habitat. They are *viz. Ceropegia attenuata, Dendrophthoe trigona, Eriocaulon eurypeplon, Habenaria grandifloriformis, Hyphaene dichotoma, Maytenus rothiana, Mammea suriga, Miliusa tomentosa, Osbeckia truncata, Pycreus malabaricus, Rivea ornata, Rotala malampuzhensis* and *Theriophonum dalzellii.*

Contribution towards *ex-situ* Conservation

In addition to the above it was also found that NBGT is supporting *ex-situ* conservation of *ca.* 900 species (including both native and exotic) belonging to *ca.* 465 genera and 100 families. Out of which 61 species (16 herbs, 8 shrubs and 37 trees) are either rare or threatened (Figure 5.2, Table 5.4). NBGT has also played a vital role in

Figure 5.1: Ex-situ Conservation of Few Endemic Plants at National Botanic Garden, Trombay.

Table 5.3: List of Endemic Plants Occurring at National Botanic Garden, Trombay

Family	*Botanical Name*	*Habit*	*Habitat*	*Flowering*
Acanthaceae	*Barleria montana* Nees	H	T	Sep.–Nov.
Acanthaceae	*Cynarospermum asperrimum* (Nees) Vollesen	H	T	Oct.–Dec.
Acanthaceae	*Ecbolium ligustrinum* (Vahl) Vollesen	H	T	Oct.–Dec.
Acanthaceae	*Eranthemum roseum* (Vahl) R.Br.	H	T	Nov.–Apr.
Acanthaceae	*Haplanthus tentaculatus* Nees	H	T	Dec.–May
Acanthaceae	*Hemigraphis latebrosa* (Roth) Nees	H	T	Nov.–Mar.
Acanthaceae	*Hygrophila serpyllum* T.Anderson	H	T	Dec.–Feb.
Acanthaceae	*Lepidagathis cristata* Willd.	H	T	Oct.–Mar.
Acanthaceae	*Neuracanthus sphaerostachys* Dalzell	H	T	Aug.–Dec.
Acanthaceae	*Rungia elegans* Dalzell and A.Gibson	H	T	Sep.–Feb.
Acanthaceae	*Strobilanthes callosa* Nees	S	T	Sep.–Nov.
Annonaceae	*Miliusa tomentosa* (Roxb.) Finet and Gagnep.	T	T	Apr.
Apocynaceae	*Ceropegia attenuata* Hook.	H	T	Aug.–Sep.
Apocynaceae	*Ceropegia hirsuta* Wight and Arn.	HC	T	Jul.–Aug.
Apocynaceae	*Cosmostigma cordatum* (Poir.) M.R.Almeida	HC	T	Aug.–Sep.
Apocynaceae	*Cryptolepis dubia* (Burm.f.) M.R.Almeida	HC	T	Apr.–Nov.
Apocynaceae	*Tylophora dalzellii* Hook.f.	HC	T	Aug.–Dec.
Araceae	*Amorphophallus commutatus* (Schott) Engl.	H	T	May–Jun.
Araceae	*Theriophonum dalzellii* Schott	H	T	Jul.–Aug.
Arecaceae	*Hyphaene dichotoma* (White) Furtado	T	T	Jan.–Mar.
Asparagaceae	*Dipcadi montanum* (Dalzell) Baker	H	T	Apr.–Jun.
Asteraceae	*Blumea eriantha* DC.	H	T	Nov.–May
Asteraceae	*Blumea malcolmii* Hook.f.	H	T	Nov.–May
Asteraceae	*Tricholepis glaberrima* DC.	H	T	Oct.–Jan.
Bignoniaceae	*Heterophragma quadriloculare* (Roxb.) K.Schum.	T	T	Feb.–Apr.
Boraginaceae	*Trichodesma inaequale* Edgew.	H	T	Aug.–Sep.
Capparaceae	*Cadaba fruticosa* (L.) Druce	L	T	Nov.–Mar.
Celastraceae	*Maytenus rothiana* Lobr.-Callen	S	T	Apr.–Jun.
Clusiaceae	*Mammea suriga* (Buch.-Ham. ex Roxb.) Kosterm.	T	T	Feb.–Apr.
Commelinaceae	*Murdannia semiteres* (Dalzell) Santapau	H	T/L	Jul.–Oct.
Commelinaceae	*Murdannia versicolor* (Dalzell) G.Brückn.	H	T	Oct.–Nov.
Convolvulaceae	*Argyreia pilosa* Wight and Arn.	HC	T	Sept.–Nov.
Convolvulaceae	*Argyreia sericea* Dalzell	HC	T	Aug.–Feb.
Convolvulaceae	*Ipomoea barlerioides* (Choisy) Benth. ex C.B. Clarke	HC	T	Sep.–Oct.
Convolvulaceae	*Rivea ornata* Choisy	HC	T	Jul.–Sep.
Cyperaceae	*Pycreus malabaricus* C.B.Clarke	H	T	Sep.–Nov.

Family	Botanical Name	Habit	Habitat	Flowering
Eriocaulaceae	*Eriocaulon eurypeplon* Körn.	H	T/L	Jul.–Aug.
Eriocaulaceae	*Eriocaulon lanceolatum* Miq. ex Körn.	H	T	Oct.–Nov.
Euphorbiaceae	*Acalypha malabarica* Müll.Arg.	H	T	Jul.–Aug.
Euphorbiaceae	*Bridelia hamiltoniana* Wall. ex Müll.Arg.	S	T	Aug.–Nov.
Euphorbiaceae	*Mallotus polycarpus* (Benth.) Kulju and Welzen	S	T	Feb.–Mar.
Fabaceae	*Acacia chundra* (Rottler) Willd.	T	T	Jun.–Jul.
Fabaceae	*Alysicarpus hamosus* Edgew.	H	T	Aug.–Sep.
Fabaceae	*Clitoria annua* J.Graham	H	T	Aug.–Nov.
Fabaceae	*Crotalaria filipes* Benth. var. *filipes*	H	T	Jul.–Feb.
Fabaceae	*Crotalaria filipes* var. *trichophora* (Baker) T.Cooke	H	T	Jul.–Feb.
Fabaceae	*Crotalaria leptostachya* Benth.	H	T	Oct.–Dec.
Fabaceae	*Crotalaria lutescens* Dalzell	H	T	Nov.–Dec.
Fabaceae	*Indigofera glandulosa* Wendl.	H	T	Sep.–Feb.
Fabaceae	*Neptunia triquetra* (Vahl) Benth.	H	T	Aug.–Jan.
Gentianaceae	*Exacum pumilum* Griseb.	H	T	Aug.–Dec.
Lamiaceae	*Anisomeles heyneana* Benth.	H	T	Jun.–Nov.
Lamiaceae	*Pogostemon purpurascens* Dalzell	H	T	Oct.–Jan.
Loranthaceae	*Dendrophthoe trigona* (Wight and Arn.) Danser ex Santapau	S	P	Dec.–Feb.
Lythraceae	*Lagerstroemia lanceolata* Wall.	T	T	Apr.–Jun.
Lythraceae	*Lagerstroemia parviflora* Roxb.	T	T	Mar.–Jun.
Lythraceae	*Rotala malampuzhensis* R.V.Nair	H	A	Aug.–Sep.
Malpighiaceae	*Aspidopterys cordata* (B.Heyne ex Wall.) A.Juss.	L	T	Sep.–Nov.
Malvaceae	*Firmiana subglabra* (V.Abraham and Dutt) Kosterm.	T	T	Apr.–May
Melastomataceae	*Osbeckia truncata* D.Don ex Wight and Arn.	H	T	Aug.–Sep.
Oleaceae	*Jasminum malabaricum* Wight	L/S	T	Feb.–May
Orchidaceae	*Habenaria grandifloriformis* Blatt. and McCann	H	T	Jul.–Aug.
Poaceae	*Arundinella metzii* Hochst. ex Miq.	H	T	Oct.–Feb.
Poaceae	*Sehima sulcatum* (Hack.) A.Camus	H	T	Oct.–Dec.
Rhamnaceae	*Ziziphus caracutta* Buch.-Ham. ex Roxb.	T	T	Jul.–Aug.
Rhamnaceae	*Ziziphus glaberrima* (Sedgw.) Santapau	S	T	Mar.–Apr.
Rhamnaceae	*Ziziphus xylopyrus* (Retz.) Willd.	T	T	Jul.–Aug.
Rubiaceae	*Anotis foetida* Hook.f.	H	T	Jul.–Sep.
Vitaceae	*Cissus woodrowii* (Stapf ex Cooke) Santapau	S	T	May–Jul.
Zingiberaceae	*Curcuma inodora* Blatt.	H	T	Apr.–Jun.

Habit: H = Herb, HC = Herbaceous Climber, L = Liana, S = Shrub, T= Tree; Habitat: A = Aquatic, L = Lithophyitc, P = Parasitic, T = Terrestrial.

Figure 5.2: Ex-situ Conservation of Few Rare Plants at National Botanic Garden, Trombay.

Table 5.4: *Ex-situ* Conservation of Few Rare and Threatened Plants at National Botanic Garden, Trombay

Family	*Botanical Name*	*Habit*	*Habitat*	*Flowering*	*IUCN Status*
Anacardiaceae	*Pleiogynium timoriense* (A. DC.) Leenh.	T	T	Mar.–Apr.	-
Apocynaceae	*Alstonia macrophylla* Wall. ex G.Don	T	T	Jan.–Feb.	LC
Apocynaceae	*Frerea indica* Dalzell	H	T	Sep.–Oct.	EN
Arecaceae	*Adonidia merrillii* (Becc.) Becc.	T	T	Sep.–Oct.	NT
Arecaceae	*Aiphanes minima* (Gaertn.) Burret	T	T	-	-
Arecaceae	*Arenga obtusifolia* Mart.	T	T	-	-
Arecaceae	*Attalea cohune* Mart.	T	T	-	-
Arecaceae	*Bismarckia nobilis* Hildebr. and H.Wendl.	T	T	-	LC
Arecaceae	*Coccothrinax argentea* (Lodd. ex Schult. and Schult.f.) Sarg. ex Becc.	T	T	-	
Arecaceae	*Coccothrinax barbadensis* (Lodd. ex Mart.) Becc.	T	T	-	-
Arecaceae	*Corypha utan* Lam.	T	T	-	LC
Arecaceae	*Dictyosperma album* (Bory) Scheff.	T	T	-	-
Arecaceae	*Dypsis lutescens* (H.Wendl.) Beentje and J.Dransf.	S	T	Mar.–May	NT
Arecaceae	*Dypsis madagascariensis* (Becc.) Beentje and J.Dransf.	T	T	Apr.–May	LC
Arecaceae	*Latania lontaroides* (Gaertn.) H.E.Moore	T	T	-	EN
Arecaceae	*Livistona chinensis* (Jacq.) R.Br. ex Mart.	T	T	-	-
Arecaceae	*Ptychosperma elegans* (R.Br.) Blume	S	T	Apr.–May	-
Arecaceae	*Ptychosperma macarthurii* (H.Wendl. ex H.J.Veitch) H.Wendl. ex Hook.f.	S	T	Apr.–May	-
Arecaceae	*Sabal minor* (Jacq.) Pers.	S	T	Jan.–Feb.	-
Arecaceae	*Sabal palmetto* (Walter) Lodd. ex Schult. and Schult.f.	T	T	-	-
Arecaceae	*Saribus rotundifolius* (Lam.) Blume	T	T	-	-
Arecaceae	*Wodyetia bifurcata* A.K.Irvine	T	T	-	LC
Bignoniaceae	*Dolichandrone spathacea* (L.f.) Seem.	T	T	Jun.–Jul.	LC
Clusiaceae	*Garcinia indica* (Thouars) Choisy	T	T	Dec.–Feb.	VU
Combretaceae	*Terminalia tomentosa Wight and Arn.*	T	T	May.–Jun.	-
Cycadaceae	*Cycas circinalis* L.	T	T	Feb.–Mar.	EN
Cycadaceae	*Cycas revoluta* Thunb.	T	T	-	LC
Elaeocarpaceae	*Elaeocarpus grandiflorus* Sm.	S	T	Apr.	-
Fabaceae	*Pterocarpus indicus* Willd.	T	T	Mar.–Apr.	VU
Fabaceae	*Saraca asoca* (Roxb.) De Wilde	T	T	Mar.–May	VU

Family	*Botanical Name*	*Habit*	*Habitat*	*Flowering*	*IUCN Status*
Lauraceae	*Cinnamomum verum* J.Presl	T	T	Dec.–Jan.	-
Loganiaceae	*Strychnos nux-vomica* L.	T	T	Mar.–Apr.	-
Malvaceae	*Berrya cordifolia* (Willd.) Burret	T	T	May.–Jun.	-
Malvaceae	*Berrya javanica* (Turcz.) Burret	T	T	Oct.–Dec.	-
Meliaceae	*Khaya senegalensis* (Desv.) A.Juss.	T	T	Apr.–May	VU
Meliaceae	*Swietenia macrophylla* King	T	T	Apr.–May	VU
Meliaceae	*Swietenia mahogani* L.	T	T	Feb.–May	-
Orchidaceae	*Acampe carinata* (Griff.) Panigrahi	H	E	Dec.–Jan.	-
Orchidaceae	*Acampe rigida* (Buch.-Ham. ex Sm.) P.F.Hunt	H	E	Dec.–Jan.	-
Orchidaceae	*Aerides multiflorum* Roxb.	H	E	Apr.–Jul.	-
Orchidaceae	*Aerides odorata* Lour.	H	E	May–Jul.	-
Orchidaceae	*Arundina graminifolia* var. *revoluta* (Hook.f.) A.L.Lamb	H	T	Oct.–Nov.	-
Orchidaceae	*Bulbophyllum crassipes* Hook.f.	H	E	Sep.–Oct.	-
Orchidaceae	*Coelogyne nitida* (Wall. ex D.Don) Lindl.	H	E	May	-
Orchidaceae	*Cymbidium aloifolium* (L.) Sw.	H	E	Apr.–Jun.	-
Orchidaceae	*Dendrobium aphyllum* (Roxb.) C.E.C.Fisch.	H	E	Feb.–Mar.	LC
Orchidaceae	*Dendrobium transparens* Wall. ex Lindl.	H	E	Mar.–May	-
Orchidaceae	*Eria lasiopetala* (Willd.) Ormerod	H	E	Jan.–Mar.	-
Orchidaceae	*Pelatantheria insectifera* (Rchb.f.) Ridl.	H	E	Dec.–Jan.	-
Orchidaceae	*Phaius tankervilleae* (Banks) Blume	H	T	Mar.–May	-
Orchidaceae	*Pholidota articulata* Lindl.	H	E	Apr.	-
Orchidaceae	*Pholidota imbricata* Lindl.	H	E	May.–Aug.	-
Paulowniaceae	*Paulownia tomentosa* Steud.	T	T	May.–Jun.	-
Platanaceae	*Platanus orientalis* L.	T	T	-	LC
Santalaceae	*Santalum album* L.	T	T	Dec.–Jan.	VU
Sapindaceae	*Melicoccus bijugatus* Jacq.	T	T	Mar.	-
Sapotaceae	*Madhuca longifolia* (J.Koenig ex L.) J.F.Macbr. var. *longifolia*	T	T	Mar.–May	-
Sapotaceae	*Manilkara kauki* (L.) Dubard	T	T	Jun.–Jul.	-
Solanaceae	*Withania somnifera* (L.) Dunal	S	T	Feb.–Mar.	-
Zamiaceae	*Zamia angustifolia* Jacq.	S	T	-	VU
Zamiaceae	*Zamia furfuracea* L.f. ex Aiton	S	T	-	EN

Habit: H = Herb, S = Shrub, T= Tree; Habitat: E = Epiphytic, T = Terrestrial; IUCN Status: EN = Endangered, LC = Least Concern, NT = Near Threatened, VU = Vulnerable.

Figure 5.3: Few Rare Ornamental Trees Conserved at National Botanic Garden, Trombay.

ex-situ conservation of tree species in the form of arboreta, where *ca.* 170 species of trees belonging to 48 families have been conserved. Among notable tree collections following forty ornamental trees are of rare occurrence and are hardly seen in any botanic gardens. They are *viz. Acacia mangium* Willd., *Adansonia digitata* L., *Averrhoa bilimbi* L., *Bonellia macrocarpa* (Cav.) B.Ståhl and Källersjö, *Bauhinia monandra* Kurz, *Ceiba insignis* (Kunth) P.E.Gibbs and Semir, *Brownea grandiceps* Jacq., *Cassia roxburghii* DC., *Calophyllum inophyllum* L., *Cerbera odollam* Gaertn., *C. manghas* L., *Chrysophyllum cainito* L., *Coccoloba uvifera* (L.) L., *Colvillea racemosa* Bojer, *Castanospermum australe* A.Cunn. and C.Fraser, *Clusia rosea* Jacq., *Cordia subcordata* Lam., *Corymbia citriodora* (Hook.) K.D.Hill and L.A.S.Johnson, *Couroupita guianensis* Aubl, *Crateva adansonii* subsp. *odora* (Buch.-Ham.) Jacobs, *C. nurvala* Buch.-Ham., *Eucalyptus platyphylla* F.Muell., *Ficus krishnae* C.DC., *Filicium decipiens* (Wight and Arn.) Thwaites, *Guazuma ulmifolia* Lam., *Gustavia augusta* L., *Handroanthus impetiginosus* (Mart. ex DC.) Mattos, *Kigelia africana* (Lam.) Benth., *Lepisanthes tetraphylla* (Vahl) Radlk., *Markhamia lutea* (Benth.) K.Schum., *Melaleuca leucadendra* (L.) L., *M. quinquenervia* (Cav.) S.T.Blake, *Myroxylon balsamum* (L.) Harms, *Pseudobombax ellipticum* (Kunth) Dugand, *Pterygota alata* (Roxb.) R.Br. 'Diversifolia', *Paulownia tomentosa* Steud, *Spathodea campanulata* P.Beauv., *Tabebuia aurea* (Silva Manso) Benth. and Hook.f. ex S.Moore, *T. berteroi* (DC.) Britton, *Terminalia muelleri* Benth., *Vernicia fordii* (Hemsl.) Airy Shaw and *Vitex megapotamica* (Spreng.) Moldenke (Figure 5.3).

Besides above, it is worthwhile to mention here that this garden is separately maintaining a germplasm collection *ca.* 60 *Canna* cultivars and *ca.* 140 cultivars of genus *Bougainvillea* for various ornamental purposes and sustainable gardening. In addition to the above, NBGT has also established a rose garden in its campus especially dedicated for germplasm conservation of Indian rose cultivars. So far *ca.* 150 Indian rose cultivars have been conserved in this garden.

Acknowledgements

Authors are thankful to Shri K.N. Vyas, Director, BARC, for providing facilities. We also extend our sincere thanks to Shri R. Narasimhan, Head, A and SED and Shri H. Mishra AD, ESG, for their support. Authors are also thankful to Dr. P. Suprasanna SO/H and Apex Coordinator, XII Plan Project – NABTD: *"Nuclear agriculture for sustainability and societal benefits"* for encouragements.

References

Abraham V. 1997. *The flora of Trombay*. Bhabha Atomic Research Centre, Mumbai.

Abraham V. 1998. *Biodiversity at BARC Trombay*. Bhabha Atomic Research Centre, Mumbai.

Almeida M.R. 1996-2010. *Flora of Maharashtra-Vol. I-VI.* Blatter Herbarium, St. Xavier's College, Mumbai.

Arisdason W. and Lakshminarasimhan, P. 2016. Status of plant diversity in India: an overview. Available from: http://www.bsienvis.nic.in/Database/Status_of_Plant_Diver-sity_in_India_17566.aspx. Accessed on 26 Apr. 2016

Ashton P.S. 1988. Conservation of Biological Diversity in Botanical Gardens. In: Wilson, E.O. and Peter, M. (eds), *Biodiversity*. National Academy Press, Washington, D.C. pp. 269–278.

Bapat V.A., Dixit G.B. and Yadav S.R. 2012. Plant biodiversity conservation and role of botanists. *Current Science* 102(10):1366–1369.

BGCI, 2016. Botanic Gardens Conservation International: History. Available from https://www.bgci.org/resources/history/Accessed on 10th May 2016.

CBD, 2016. Convention on Biological Diversity: Botanic Gardens recorded in India. Available from https://www.cbd.int/doc/world/in/in-ex-bg-en.pdf Accessed on 10th May 2016.

Cooke T. 1901-1908. *The Flora of the presidency of Bombay. Vol. I and II.* Taylor and Francis, London.

de Vos J.M., Joppa L.N., Gittleman J.L., Stephens P.R. and Pimm S.L. 2014 Estimating the Normal Background Rate of Species Extinction. *Conservation Biology* 29(2): 452–462.

Gomez-Mejia, A. *et al.*, 2006. *The Gran Canaria Declaration II on Climate Change and Plant Conservation*. Jardin Botanico Canarion and Botanic Gardens Conservation International.

Hooker J.D. 1872-1897. *The flora of British India. Vol. I-VII.* L. Reeve and Co., Ltd., Kent.

IUCN, 2001. IUCN Red List Categories and Criteria: Version 3.1. IUCN Species Survival Commission. World Conservation Union, Gland, Switzerland and Cambridge, United Kingdom.

Jain S.K. and Rao R.R. 1977. *A Handbook of Field and Herbarium Methods*. Today and Tomorrows Printers and Publishers, New Delhi.

Mishra D.K. and Singh N.P. 2001. *Endemic and Threatened Flowering Plants of Maharashtra*. Botanical Survey of India, Calcutta.

Nayar M.P and Sastry A.R.K. (eds) 1987-1990. *Red Data Books of Indian Plants. Vol. I-III*. Botanical Survey of India, Calcutta.

Ogale, V.K. 2010. *Heritage trees of Trombay*. Bhabha Atomic Research Centre, Mumbai.

Rao C.K., Geetha B.L. and Suresh G. 2003. *Red List of Threatened Vascualr plants Species in India*. ENVIS, Botanical Survey of India, Kolkata.

Sharma B.D., Karthikeyan S. and Singh N.P. 1996. *Flora of Maharashtra State (Monocot.)*. Botanical Survey of India, Kolkata.

Singh N.P. and Karthikeyan S. 2000. *Flora of Maharashtra State (Dicotyledones) Vol. 1*. (Ranunculaceae to Rhizophoraceae). Botanical Survey of India, Kolkata.

Singh N.P., Lakshminarasimhan P., Karthikeyan S. and Prasanna P.V. 2001. *Flora of Maharashtra State (Dicotyledones) Vol. 2 (Combretaceae to Ceratophyllaceae)*. Botanical Survey of India, Kolkata.

Singh P. and Dash S.S. 2015. *Plant Discoveries 2014 – New Genera, Species and New Records*. Botanical Survey of India, Kolkata.

Singh P., Karthigeyan K., Lakshminarasimhan and Dash S.S. 2015. *Endemic Vascular Plants of India*. Botanical Survey of India, Kolkata.

Chapter 6

Conservation Profile of the Bhagalpur University Botanical Garden, Bhagalpur (Bihar)

T.K. Pan[1] *and C.B. Singh*[2]

University Department of Botany, T.M. Bhagalpur University, Bhagalpur – 812 007, Bihar
E-mail: [1]tkpan.botany@gmail.com, [2]chandra.bhanu.singh31@gmail.com

Abstract

The global biological extinction is a challenge for the environmentalists. The magnitude of its causal factors has widened to the extent that one plant species is becoming extinct everyday on an average. Several measures are adopted to check this serious menace. Of them, ex-situ conservation in the botanical gardens is proving to be the most effective one. The botanic gardens are playing very significant role in conservation, multiplication and rehabilitation of threatened (the collective term used for rare, endangered, vulnerable and threatened) and endemic plants. The Bhagalpur University Botanical Garden, Bhagalpur (Bihar) is one of them. At present, it has a living collection of more than 1200 plant species containing pteridophytes (16), gymnosperms (21) and angiosperms (1166). The rich profile of garden encompasses 46 threatened (30 rare, 5 endangered, 2 vulnerable and 9 threatened) and 1 endemic species. These threatened and endemic plants are economically useful and comprised of 3 gymnosperms and 44 angiosperms represented by 39 dicots and 5 monocots. Most of the plants are grown well in the garden soil but some species (Frerea indica L., Turnera ulmifolia L.) as pot plants or marshy plants in water tanks (Acorus calamus L.). These plants are propagated by seeds or/and vegetative propagules (bulbils, tubers, stem cuttings, etc.), their combinations proving effective means in certain cases. The proper maintenance of a large number of economic plants and emphasis on conservation cum multiplication of threatened and endemic plants may enable the garden to become an excellent gene pool conservation centre in Eastern India.

Keywords: *Botanic gardens, Ex-situ conservation, Threatened plants.*

Introduction

The Indian subcontinent is one of the mega gene pool centres of the world. It contains very rich plant diversity with an estimated 50,000 plant species including 17,000 flowering plants. Paradoxically, 10-15 per cent of the Indian flowering plants are under various degrees of threat. The data on threatening of plants may reach up to 25 per cent by the turn of the century unless proper conservation measures are taken. Realizing it, many threatened and endemic plants are being conserved in *ex-situ* condition in different botanical gardens of our country. The Bhagalpur University Botanical Garden, Bhagalpur (Bihar) is one of them. It is one of the best botanical gardens of Eastern India. Since its establishment in 1970, it is playing a significant role in the conservation of threatened and endemic flowering plants. Regularly, its conservation profile is enriching due to addition of some new plants with any degree of threat. At present, it is conserving 47 threatened and endemic flowering plants which are documented in this chapter.

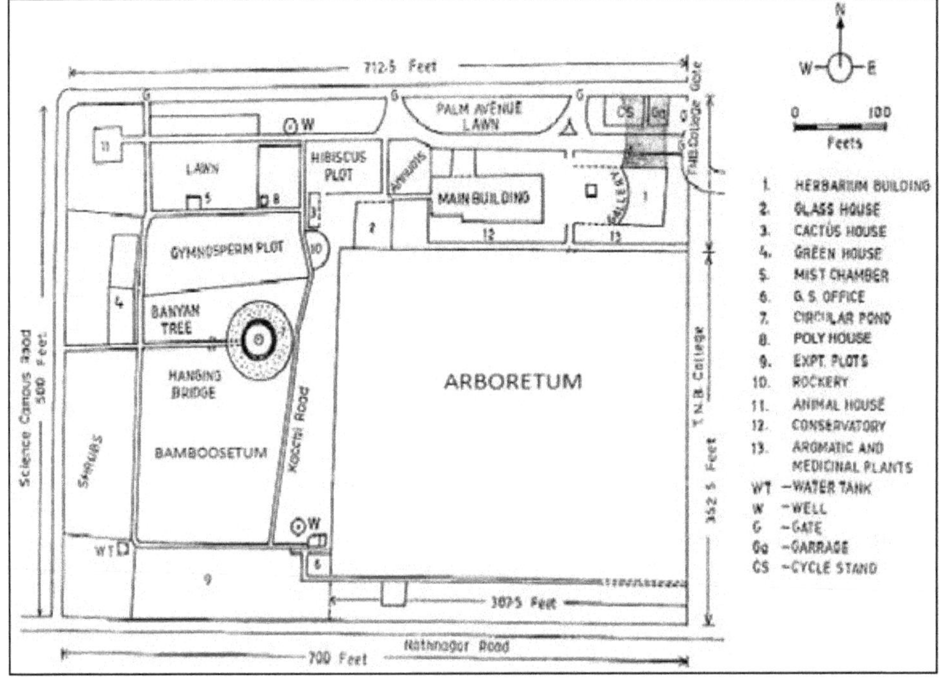

Figure 6.1: Layout Map of Bhagalpur University Botanical Garden, Bhagalpur (Bihar).

Topography and Climate

The Bhagalpur University Botanical Garden, Bhagalpur (Bihar) is situated at 25°14′26″NL and 86°56′53″EL, and at an altitude of 52m above mean sea level. It is spread over a plain area of more than eight acres (Figure 6.1). It is characterized by hot- dry summer (March to mid-June), warm-moist rainy (mid-June to October) and cold-sub-humid winter (November to February) seasons. The annual rainfall varies from 665.0 mm to 1364.7 mm. The monthly temperature ranges between

7.5°C (January) to 35.9°C (April/May). Overall January remains coolest whereas May hottest in the year. The monthly relative humidity fluctuates from 38.2 per cent (March) to 94.6 per cent (December). The wind velocity varies much from its minimum 2.2km/h in November to maximum 8.1km/h in July. More or less, the climatic condition favours the growth of plants in their respective season of the year.

Soil and Infrastructure

The garden soil is alluvium, sandy loam, neutral to slightly alkaline in reaction (pH: 7.0-7.4) and has high water holding capacity (47 per cent).Its temperature ranges from 25-35°C and moisture content in between 7.1 -24.7 per cent during the different months of the year. Similarly, the level of its nutrients shows temporal fluctuation in particular range, *viz.*, organic carbon: 0.57-0.71 per cent, total nitrogen 0.05 – 0.06 per cent, available phosphorous: 120-186ppm and exchangeable potassium: 120-150ppm. The very narrow range of C/N ratio (11.4 -11.8) indicates that the soil maintains its fertility throughout the year and thus supports the growth of plants in the garden.

The infrastructure of the garden includes conservatory, mist chamber, poly house, cactus house, green house, herbarium, aromatic and medicinal plants garden, palm garden, gymnosperm garden, aquatic garden (pond and water tanks), bamboosetum, arboretum, *etc.* which are conducive to goodness of the garden.

Conservation Profile

Since the time of its establishment, the main objective of Bhagalpur University Botanical Garden, Bhagalpur (Bihar) is conservation of threatened and endemic flowering plants of erstwhile Bihar state (now divided into two states – Bihar and Jharkhand). For this purpose, the forests and other natural habitats of these two states as well as adjoining areas of neighbouring provinces are visited frequently to collect the plants and to observe their threat status on spot. The flowering plants facing various degrees of threat are collected from the diverse places, identified properly with the help of different floras or monographs, assessed to arrive their threat status (rare, endangered, vulnerable and threatened) as per Red Data Book of Indian Plants (BSI,1987-1990)/IUCN (2015) and brought to *ex- situ* conservation. The herbarium sheets of these plants are housed in its own Herbarium which is now known as "Herbarium of University Department of Botany, Tilka Manjhi Bhagalpur University, Bhagalpur (Bihar)" having 4000 species of angiosperms (numerically about 95 per cent), gymnosperms and ferns collected mainly from Bihar, Jharkhand and Nepal. The gradual introduction of threatened and endemic plants in the garden is furthering its conservation list. At present, forty seven threatened and endemic plants including four from neighbouring states (West Bengal, Odisha and Uttar Pradesh) are under conservation in the garden. All these plants are arranged in alphabetical order with respect to their botanical name, placed in the respective family and assigned to the specific threat category (Table 6.1) for documentation of conservation profile of the garden.

The threatened and endemic flowering plants conserved in the garden are comprised of 3 gymnosperms (*Cycas pectinata* Griff., *Cycas rumphii* Miq. and *Ginkgo biloba* L.) and 44 angiosperms. Among the angiosperms, only five species, *viz.*, *Acorus*

Table 6.1: Threatened and Endemic Flowering Plants under Conservation in Bhagalpur University Botanical Garden, Bhagalpur (Bihar)

Botanical Name	*Collection Place*	*Habit*	*Threat Status*
Acorus calamus L. (Araceae)	Motijharna, Rajmahal hills (Jharkhand)	Herb	Threatened
Aegle marmelos (L.) Corr. (Rutaceae)	Motijharna, Rajmahal hills (Jharkhand)	Tree	Rare
Albizia lebbeck (L.) Benth. (Mimosaceae)	Silingi, Santhal Pargana forest (Jharkhand)	Tree	Rare
Albizia procera Benth.(Mimosaceae)	Silingi, Santhal Pargana forest (Jharkhand)	Tree	Rare
Aphanamixis polystachya (Wall.) Park. (Meliaceae)	Saranda forest (Jharkhand)	Tree	Rare
Asparagus racemosus Willd. (Liliaceae)	Motijharna, Rajmahal hills (Jharkhand)	Climber	Rare
Bauhinia vahlii Wt. and Arn. (Caesalpiniaceae)	Saranda forest (Jharkhand)	Climber	Rare
Bixa orellana L.(Bixaceae)	Saranda forest (Jharkhand)	Tree	Rare
Boswellia serrata Roxb. ex Colebr. (Bursaraceae)	Saranda forest (Jharkhand)	Tree	Rare
Canavalia gladiata (Jacq.) DC. (Fabaceae)	Bhatoria, Bhagalpur (Bihar)	Climber	Rare
Clerodendrum serratum (L.) Moon (Verbenaceae)	Jitpur, Santhal Pargana forest (Jharkhand)	Shrub	Endangered
Cochlospermum religiosum (L.) Alston (Cochlospermaceae)	Saranda forest (Jharkhand)	Tree	Rare
Coffea bengalensis Roxb.(Rubiaceae)	Motijharna, Rajmahal hills (Jharkhand)	Shrub	Rare
Commiphora roxburghii (Wt. and Arn.) Engl. var. *serratifolia* Haines (Bursaraceae)	Mayganj, Bhagalpur (Bihar)	Tree	Endemic
Costus speciosus (Koen.) Sm. (Zingiberaceae)	Motijharna, Rajmahal hills (Jharkhand)	Herb	Threatened
Cycas pectinata Griff. (Cycadaceae)	Someshwar hills, West Champaran forest (Bihar)	Tree	Vulnerable
Cycas rumphii Miq. (Cycadaceae)	Pant Botanical Garden, Allahabad University (U.P.)	Tree	Endangered
Dalbergia lanceolaria L. f. (Fabaceae)	Kathikund, Santhal Pargana forest(Jharkhand)	Tree	Rare
Dalbergia latifolia Roxb. (Fabaceae)	Gopikandar, Santhal Pargana forest (Jharkhand)	Tree	Rare
Dillenia pentagyna Roxb. (Dilleniaceae)	Silingi, Santhal Pargana forest (Jharkhand)	Tree	Rare
Diospyros cordifolia Roxb. (Ebenaceae)	Motijharna, Rajmahal hills (Jharkhand)	Tree	Rare
Diospyros discolor Willd.(Ebenaceae)	Ranchi (Jharkhand)	Tree	Threatened
Diospyros malabarica (Desr.) Kostel (Ebenaceae)	Motijharna, Rajmahal hills (Jharkhand)	Tree	Threatened

Botanical Name	*Collection Place*	*Habit*	*Threat Status*
Diospyros montana Roxb. (Ebenaceae)	Motijharna, Rajmahal hills (Jharkhand)	Tree	Threatened
Emblica officinalis Gaertn. (Euphorbiaceae)	Banjhi, Santhal Pargana forest (Jharkhand)	Tree	Rare
Ficus rumphii Blume (Moraceae)	Motijharna, Rajmahal hills (Jharkhand)	Tree	Threatened
Frerea indica L. (Asclepiadaceae)	BSI, Kolkata (West Bengal)	Herb	Endangered
Ginkgo biloba L.(Ginkgoaceae)	Kalimpong, Darjiling (West Bengal)	Tree	Endangered
Gloriosa superba L.(Liliaceae)	Sundarpahari, Santhal Pargana forest (Jharkhand)	Climber	Endangered
Ixora undulata Roxb. (Rubiaceae)	Motijharna, Rajmahal hills (Jharkhand)	Tree	Rare
Oroxylum indicum (L.) Vent. (Bignoniaceae)	Sundarpahari, Santhal Pargana forest (Jharkhand)	Tree	Rare
Polyalthia suberosa (Roxb.) Thw. (Annonacaae)	Motijharna, Rajmahal hills (Jharkhand)	Shrub	Rare
Pterocarpus marsupium Roxb. (Fabaceae)	Gua, Saranda forest (Jharkhand)	Tree	Vulnerable
Rauvolfia serpentina (L.) Benth. ex Kurz (Apocynaceae)	Kathikund, Santhal Pargana forest (Jharkhand)	Under Shrub	Threatened
Rosa involucrata Roxb. (Rosaceae)	Mirzachowki, Rajmahal hills (Jharkhand)	Shrub	Rare
Semecarpus anacardium L.f. (Anacardiaceae)	Gopikandar, Santhal Pargana forest (Jharkhand)	Shrub	Rare
Stemona tuberose Lour. (Stemonaceae)	Ganjam forest (Odisha)	Climber	Rare
Sterculia urens Roxb. (Sterculiaceae)	Motijharna, Rajmahal hills (Jharkhand)	Tree	Rare
Strychnos nux-vomica (L.) Benth. ex Kurz (Loganiaceae)	Saranda forest (Jharkhand)	Tree	Threatened
Suregada multiflorum (Juss.) Baill. (Euphorbiaceae)	Silingi, Santhal Pargana forest (Jharkhand)	Tree	Rare
Tacca leontopetaloides (L.) Kuntz. (Taccaceae)	Maharajpur, Sahebganj (Jharkhand)	Herb	Rare
Tephrosia candida Wall. (Fabaceae)	Barajamda,Saranda forest (Jharkhand)	Shrub	Threatened
Terminalia bellirica (Gaertn.) Roxb. (Combretaceae)	Kathikund, Santhal Pargana forest (Jharkhand)	Tree	Rare
Terminalia chebula Retz. (Combretaceae)	Kathikund, Santhal Pargana forest (Jharkhand)	Tree	Rare
Turnera ulmifolia L.(Turneraceae)	Barajamda, Saranda forest (Jharkhand)	Shrub	Rare
Uraria picta (Jacq.) Desv. ex DC. (Fabaceae)	Silingi, Santhal Pargana forest (Jharkhand)	Herb	Rare
Vitex peduncularis Wall. ex Schauz (Verbenaceae)	Sundarpahari, Santhal Pargana forest (Jharkhand)	Small tree	Rare

calamus L., *Asparagus racemosus* Willd., *Costus speciosus* (Koen.) Sm., *Gloriosa superba* L. and *Tacca leontopetaloides* (L.) Kuntze represent monocotyledons and the rest ones belong to dicot families. The threatened plants with different threat status are maximum in number in Fabaceae. These plants are economically useful for timber – *Albizia lebbeck* (L.) Benth., *Albizia procera* Benth., *Aphanamixis polystachya* (Wall.) Park., *Boswellia serrata* Roxb. ex Colebr., *Dalbergia lanceolaria* L. f., *Dalbergia latifolia* Roxb., *Diospyros cordifolia* Roxb., *Diospyros discolor* Willd., *Diospyros malabarica* (Desr.) Kostel, *Diospyros montana* Roxb., *Pterocarpus marsupium* Roxb.; fuelwood – *Aegle marmelos* (L.) Corr., *Coffea bengalensis* Roxb., *Ficus rumphii* Blume, *Ixora undulata* Roxb., *Polyalthia suberosa* (Roxb.) Thw.; drugs – *Acorus calamus* L., *Aegle marmelos* (L.) Corr., *Asparagus racemosus* Willd., *Clerodendrum serratum* (L.) Moon, *Cochlospermum religiosum* (L.) Alston, *Commiphora roxburghii* (Wt. and Arn.) Engl. *var. serratifolia* Haines, *Dillenia pentagyna* Roxb., *Emblica officinalis* Gaertn., *G. superba* L., *Oroxylum indicum* (L.) Vent., *Rauvolfia serpentina* (L.) Benth. ex Kurz, *Sterculia urens* Roxb., *Strychnos nux-vomica* (L.) Benth. ex Kurz, *T. leontopetaloides* (L.) Kuntz., *Terminalia bellirica* (Gaertn.) Roxb., *Terminalia chebula* Retz., *Uraria picta* (Jacq.) Desv. ex DC., *Vitex peduncularis* Wall. ex Schauz.; *edible fruits – Emblica officinalis* Gaertn, *Ficus rumphii Blume, P. suberosa* (Roxb.) Thw.; *fruit vegetable – Canavalia gladiata* (Jacq.) DC.; *ornamental flowers – Frerea indica* L., *Rosa involucrata* Roxb., *Stemona tuberosa* Lour., *Tephrosia candida* Wall., *Turnera ulmifolia* L.; *fish poison – Suregada multiflorum* (Juss.) Baill.; *and miscellaneous value – Bauhinia vahlii* Wt. and Arn., *Bixa orellana* L., *Costus speciosus* (Koen.) Sm., *Cycas pectinata* Griff., *Cycas rumphii* Miq., *Ginkgo biloba L., Semecarpus anacardium* L.f.

The current conservation profile of the garden concerns 46 threatened (30 rare, 5 endangered, 2 vulnerable and 9 threatened) and 1 endemic species (Table 6.1, Figure 6.2). Most of them perform well in the garden soil but some species grow either only in the pots (*F. indica* L., *T. ulmifolia* L.) or water tanks (*A. calamus* L. as a marshy plant). Irrespective of the threat status, they are propagated mainly through their seeds. However, certain species are multiplied only by vegetative means like bulbils – *C. pectinata* Griff., *C. rumphii* Miq.; *tubers – S. tuberosa* Lour., *T. leontopetaloides* (L.) Kuntz.; *stem cuttings – Acorus calamus* L., *C. roxburghii* (Wt. and Arn.) Engl. *var. serratifolia* Haines, *C. speciosus* (Koen.) Sm., *D. discolor* Willd., *F. indica* L.,*G. biloba* L., *I. undulata* Roxb. The combinations of different propagules such as seeds and stem cuttings – *C. bengalensis* Roxb., *D. lanceolaria* L. f., *D. latifolia* Roxb., *E. officinalis* Gaertn., *F. rumphii* Blume, *R. serpentina* (L.) Benth. ex Kurz, *R. involucrata* Roxb., *S. anacardium* L.f., *S. multiflorum* (Juss.) Baill., *T. candida* Wall., *T. ulmifolia* L., *V. peduncularis* Wall. ex Schauz.; seeds and tubers – *A. racemosus* Willd.; seeds, stem cuttings and tubers – *G. superba* L. are rather effective in rapid multiplication of the threatened plants.

Discussion

Day-to-day, the dimension of destructive anthropogenic activities is widening to extent that a number of species of the past are either disappearing or facing various degrees of threat. For instance, the species such as *A. marmelos* (L.) Corr., *B. vahlii* Wt. and Arn., *B. serrata* Roxb. ex Colebr., *D. lanceolaria* L. f., *D. montana* Roxb., *E. officinalis* Gaertn., *O. indicum* (L.) Vent., *P. suberosa* (Roxb.) Thw., *P. marsupium*

Roxb., *S. anacardium* L.f., *S. urens* Roxb., *T. bellirica* (Gaertn.) Roxb., *T. chebula* Retz., *V. peduncularis* Wall. ex Schauz., recorded as depleting taxa in the recent past (Varma and Singh, 1998), are now threatened in their natural habitats (Table 6.1, Figure 6.2). The conservation of these plants even before the time of their declaration as threatened taxa could happen owing to rich profile of the garden which has a living collection of more than 1200 plant species distributed into different groups as pteridophytes (16), gymnosperms (21) and angiosperms (1166). In the times to come, this attribute can approve the garden as an excellent gene pool conservation centre of rare, endangered, vulnerable and threatened plant species. This would be possible due to *ex-situ* conservation of the plants going to become threatened in the future and their multiplication in the botanical garden for rehabilitation in nature as most effective strategies for conserving the plant genetic diversity (Sharma and Goel, 1994; Roy, 2003).

Acknwledgement

We are thankful to Prof. S. K. Choudhary (Head, University Department of Botany, Tilka Manjhi Bhagalpur University, Bhagalpur, Bihar) and Prof S. K. Varma (Formerly Head, University Department of Botany, Tilka Manjhi Bhagalpur University, Bhagalpur, Bihar) for their valuable co-operation and suggestions during the course of manuscript preparation.

References

BSI 1987-1990. Red Data Book of Indian Plants. Vol.I-III. Howrah.

IUCN 2015. The IUCN Red List of Threatened Species, Version 2015.4. Available at www.iucnredlist.org

Sharma S.C. and Goel A.K. 1994. Role of botanic gardens in modern times. *Indian Journal of Forestry*17 (3):230-238.

Roy R.K. 2003. Conservation of plant genetic diversity in India: The role of botanic gardens in the new millennium. *Journal of Non-Timber Forest Products*10 (1/2): 61-64.

Varma S. K. and Singh C. B. 1998. Ecological note on forest of Santhal Parganas, Bihar. In:Gupta B.K. (ed), Higher Plants of Indian Sub-continent.Vol.VIII.Bishen Singh Mahendra Pal Singh, Dehra Dun, U. P. (India), pp. 101-110.

Chapter 7

Ex-situ Conservation of Threatened Plants of India at Botanical Survey of India, Dehradun

Amber Srivastava[1]*, S.K. Srivastava*[1] *and L.R. Dangwal*[2]

[1]*Botanical Survey of India, Northern Regional Centre, Dehradun – 248 195*
[2]*Hemwati Nandan Bahuguna Garhwal University, SRT Campus, Badshahithaul, New Tehri – 249 199*

Abstract

The chapter deals with the conservation strategies applied for the ex-situ conservation of the Red listed plants of India in the botanical garden of Botanical Survey of India, Dehradun. The propagating material of the species are collected from the wild habitats and propagated in the nursery. The species are provided with possible specific habitat requirement to study the growth pattern and survival percentage in ex-situ condition.

Keywords: *Ex-situ, Conservation, Threatened, Endangered, Propagation.*

Introduction

India is one of the 19-mega diverse countries of the world and stand fourth in Asia besides harbors a unique combination of biological resources in its diverse habitats and ecosystem. Of the total floral diversity of the world, 11.4 percent is present in the Indian sub-continent. In the present state of our knowledge India has about 47,513 species of plants already identified of which about 28 per cent of the country's recorded flora is endemic to the country (Anonymous, 2014). In the present trend, unsustainable development and urbanization has led to large scale habitat degradation and destruction of threatened plant species from their narrow range of distribution. Some of the habitat specific species are facing risk of extinction in the near future while other commercially important species are over exploited from their wild habitat.

In recent years, the habitat loss due to developmental programmes, overgrazing, animal husbandry and tourism has resulted in the loss of biodiversity. Natural calamities such as floods, earthquakes and landslides also cause reduction in population and depletion of many species. Many species are extinct or on the verge of extinction before they are known for their uses (Biswas 1998; Goel 1992; Prakash and Singh 2001).

As a combined effect of the above mentioned factors there is a massive deplete in the population of the species and the species are forced towards more critical threats in their wild habitats. Since the factors are still prevailing in their natural habitat, the only way to protect the germplasm from being extinct is to conserve them through *ex-situ* conservation methods in suitable areas. As the species is under the direct vigilance in *ex-situ* conservation, it is more promising and easiest way of conservation. *In-situ* conservation of wild populations of plants is becoming more difficult day by day due to fragmentation and loss of habitats, invasion of alien species and changes in climate. *Ex-situ* conservation is an essential part of integrated conservation strategies that are aimed to conserve plant diversity in the wild (Cochrane, 2004).

According to the Global Strategy for Plant Conservation, 60 per cent of threatened plant species should be accessible in *ex-situ* collections by 2010, preferably in the country of origin. Priority should be given to the conservation of critically endangered species in their countries of origin (BGCI, 2002). Genetically representative *ex-situ* collections provide material for research and minimize impact to wild plant populations, offer potential adaptive management options for *in-situ* work and maintain stock to produce material for education, reintroduction and other activities (CPC, 1991; Guerrant *et al.*, 2004).

Besides working on the floristic documentary of country, Botanical Survey of India is also engaged in the conservation of threatened plants of the country in its various centers located in different regions of the country. Through *ex-situ* conservation of the threatened plants the institute is creating mass awareness among the people and also maintaining a germplasm collection of the threatened species of India. It also aims for the reintroduction of such threatened plants in marked suitable habitats for *in-situ* conservations practices.

Methodology

Through detailed scrutiny of the literature and herbarium the list of plants facing threats is compiled. Further field tours are conducted to locate the species in their natural habitats and also to collect the propagating materials in a sustainable manner. Application of different propagation techniques are experimented on the collected materials and a propagation protocol is prepared for each species. The propagated plants are first introduced into the botanical garden of the institute for knowing the survival rate and growth in *ex-situ* condition. Plantation of saplings in the suitable natural habitats is also practiced for *in-situ* conservation trials. Saplings are also distributed to interested individuals and institutes for promoting mass awareness and for dispersal of germplasm. Some of the species successfully conserved and propagated in the institute are detailed as follows:

1. *Catamixis baccharoides* Thomson (Asteraceae)

Vernacular Name: Vishpatra (Gaur, 1999)

Distribution: A narrow range endemic monotypic genus distributed in lower shiwalik belt of Western Himalaya. The species is reported from only nine localities in India from Haryana, Himachal Pradesh, Uttarakhand and Uttar Pradesh.

Status: Endangered

Collection and Propagation: Propagating material *viz.* seeds and stem cuttings were collected from natural habitat near Mansa Devi, Haridwar. The seeds were sown in vermiculite, sand and coconut peat. Cuttings were planted in pure sand in pots. Seed germination is recorded to be ca 45–60 per cent whereas rooting percentage of cuttings is ca 25–30 per cent.

Conservation efforts: Since the species is habitat specific and grows on steep rocky slopes, the propagated saplings are planted in pots containing sand and soil in ratio 3:2. Also, efforts are made to provide similar habitat by planting saplings on artificially created mounds made from rocks and silt. The species is surviving and propagating well in *ex-situ*.

2. *Frerea indica* Dalzell (Apocynaceae)

Vernacular Name: Shindal Makdi

Distribution: Endemic to Western ghats

Status: Endangered. Due to habitat loss, overexploitation and absence of fruiting in wild habitat, the population of the species is depleting from its wild habitats.

Collection and Propagation: The propagating materials were procured from the botanical garden of BSI, Pune. The plant is propagated through stem cuttings planted in sandy medium.

Conservation efforts: The propagated plants through stem cuttings are surviving well in *ex-situ* conservation at Dehradun. The plants also flowered well but no fruit set was found in the cultivated plants.

3. *Incarvillea emodi* (Royle ex Lindl.) Chatterjee (Bignoniaceae)

Vernacular Name: Lasu

Distribution: It is a chasmophyte herbaceous plant growing in rock crevices of mountains. The species is native to Himalayas and distributed from Afghanistan, India, Nepal and Pakistan. The species is reported from very few localities in India from Jammu and Kashmir, Himachal Pradesh and Uttarakhand.

Figure 7.1

A. *Catamixis baccharoides* Thomson, B. *Frerea indica* Dalzell, C. *Incarvillea emodi* (Royle ex Lindl.) Chatterjee, D. *Jasminum parkeri* Dunn, E. *Nepenthes khasiana* Hook.f., F. *Phlomoides superba* (Royle ex Benth.) Kamelin and Makhm., G. *Pittosporum eriocarpum* Royle, H. *Sophora mollis* (Royle) Baker.

Status: Vulnerable. Due to habitat loss, anthropogenic activities and natural disasters, the population of this species is rapidly decreasing from all the limited localities of its occurrence.

Collection and Propagation: The stem cuttings and seeds were collected from natural habitats near Devprayag, Mussoorie and Chakrata area of Uttarakhand. The collected seeds were sown on sandy medium for germination which gave ca 35 per cent germination. The cuttings treated with different concentrations of phytohormones (IBA and NAA) gave best rooting result >70 per cent planted in pure sand.

Conservation efforts: The propagated saplings were first shifted to polybags filled with pure sand and after 4 months shifted to pots and artificially created mound in the garden. Some plants are also introduced in suitable wild habitat. The plants propagated through cuttings flowered profusely in the month of March and April and also formed fruits.

4. *Jasminum parkeri* Dunn (Oleaceae)

Vernacular Name: Peeli Chameli

Distribution: The species is found among rock boulders and rocky slopes near roadsides. The species is endemic to Himachal Pradesh and is reported only from few localities of district Chamba.

Status: Endangered (Srivastava, 1985). Due to narrow range endemic and small population size which is prone to disturbance due to natural and anthropogenic factors.

Collection and Propagation: Stem cuttings and seeds were collected from Holi village near Chamba district of Himachal Pradesh. The seed germination is reported ca 45 per cent and rooting of stem cuttings is ca 70 per cent but survival percentage is reduced to 40 per cent in *ex-situ*.

Conservation efforts: The propagated saplings were planted in pots and saplings propagated through stem cuttings flowered in the next season but no seed set was found.

5. *Nepenthes khasiana* Hook.f.

Vernacular Name: Ghatparni

Distribution: Endemic to North-East India (Meghalaya,)

Status: Endangered (Saha *et al.*, IUCN, 2015). Due to overexploitation for medicinal and ornamental purposes and also due to low regeneration potential the species is facing threats in wild populations.

Collection and Propagation: Saplings and seeds were collected from Barapani, Shillong and are planted in different artificially created nutrient deficit medium. Seeds were also sown in different medium of which coco-peat gave the best germination result ca 40 per cent.

Conservation efforts: The species was planted for *ex-situ* conservation in the garden and also as botanical curiosities.

6. *Phlomoides superba* (Royle ex Benth.) Kamelin and Makhm. (Lamiaceae)

Vernacular Name: Gajar moola

Distribution: Western Himalaya [Afghanistan, Pakistan and India (Jammu and Kashmir, Himachal Pradesh and Uttarakhand)]

Status: Endangered. Due to habitat loss, over exploitation and low regeneration potential the species is depleting from all the known localities of its occurrence in India.

Collection and Propagation: The propagating material was collected from the type locality of the species *i.e.* Mohand on Dehradun-Saharanpur highway. The plant is also propagated through tissue culture.

Conservation efforts: The propagated saplings are planted in the garden for *ex-situ* conservation trial. The propagated plants flowered after 2 years in the botanical garden. Profuse flowering resulted in good amount of seed set was observed in *ex-situ* condition.

7. *Pittosporum eriocarpum* Royle (Pittosporaceae)

Vernacular Name: Meda-Thumri, Tomdi

Distribution: Endemic to Western Himalaya India (Uttarakhand and Himachal Pradesh) and Nepal.

Status: Endangered. The species is mainly facing threats due to habitat loss and low regeneration potential. Lopping for fodder by local people is also one of the reason for stunted growth.

Collection and Propagation: The propagating material *viz.* seeds, stem cuttings and root cuttings were collected from the wild habitat near Mussoorie and Sahastradhara. To break the dormancy, the seeds were treated with cold stratification which resulted in nearly 95 per cent germination.

Conservation efforts: The propagated saplings are planted in garden and also in wild to estimate the survival percentage. The survival percentage is <10 per cent in *in-situ* whereas ca 65 per cent in *ex-situ* conservation. The planted saplings flowered after four years of planting and satisfactory seed germination was observed.

8. *Sophora mollis* (Royle) Baker

Vernacular Name: Peeli Sakina

Distribution: Western Himalaya (China, India, Pakistan and Afghanistan)

Status: Endangered (Mamgain, 1999). The species is mainly used as fodder by local people which results in deplete of its population from wild habitats.

Collection and Propagation: Saplings and seeds of the species were collected from the forest area of Sahatradhara, Dehradun. The healthy viable seeds have shown a good germination rate of nearly 90 per cent. The propagated plants started flowering after 2 years of plantation in *ex-situ* condition also resulted in good seed set for further propagation.

Conservation efforts: The species is surviving well in *ex-situ* condition and was propagated through seeds and vegetative methods.

9. *Selaginella adunca* A. Br. ex Hleron (Selaginellaceae)

Vernacular Name: Sinduri

Distribution: Endemic to Indian Western Himalaya from Uttarakhand, Himachal Pradesh and Jammu and Kashmir.

Status: Endangered (IUCN, 1998). The species is mainly restricted to small patches in certain localities of its ocurrence that are prone to habitat loss in future.

Collection and Propagation: The plants are collected from wild localities of Mussoorie, Devprayag, and Chakrata. The species is propagated through suckers in the garden and planted in pots and artificially created mound.

Conservation efforts: The propagated plants are planted for *ex-situ* conservation in the garden and also introduced in suitable wild habitat for *in-situ* trials. The species is surviving well and propagating in both *ex-situ* and *in-situ* conditions.

10. *Indopiptadenia oudhensis* (Brandis) Brenan (Leguminosae)

Vernacular Name: Gainti

Distribution: Endemic to lower Shiwalik belt of Western Himalaya from Nepal and India (Uttarakhand and Uttar Pradesh)

Status: Vulnerable. Due to low regeneration potential, lopping for fodder and fuel the species is facing threats in its narrow range of occurrence.

Collection and Propagation: The seeds and saplings were collected from Suhelwa Wildlife Sanctuary, Uttar Pradesh. The seeds were sown is well drained soil medium composed of sand, soil and manure in ratio of 3:2:1. The seed germination is reported to be ca 50–60 per cent.

Conservation efforts: The propagated and collected saplings collected from wild are planted in garden for *ex-situ* conservation. The survival rate of saplings was ca 60 per cent in the garden.

A large number of plant species are facing threats in their natural habitat due to various biotic, abiotic and anthropogenic activities prevailing in their narrow range of habitat. The conservation of such threatened species is to be done by all possible means to protect them from being extinct in near future. Since the species are depleting at a fast rate from their natural habitat therefore conservation of these species in other suitable protected area is the only way of protecting the germplasm. The *ex-situ* conservation of such species in botanical gardens of scientific institutes, gardens, educational institute *etc.* is the most promising way of conservation. Under *ex-situ* conservation the species are under the direct vigilance and therefore factors affecting their growth can be easily maintained and controlled. Also the species in *ex-situ* conservation provide easy and approachable material for research work thereby decreasing the pressure on wild population.

References

Anonymous, (2014). *Plant Discoveries*, Botanical Survey of India, Kolkata. pp. 1-114.

BGCI (Botanic Gardens Conservation International). 2002. *International Agenda for Botanic Gardens in Conservation*. Botanic Gardens Conservation International, Kew, UK.

Biswas, S. 1998. Rare and threatened taxa in the forest flora of Tehri Garhwal Himalaya and strategy for their conservation. *Indian J. Forest.* 11 (3): 233-237.

Cochrane A. 2004. Western Australia's *ex situ* program for threatened species: a model integrated strategy for conservation. In: Guerrant E.O., Havens K., Maunder M. (eds) *Ex Situ Plant Conservation Supporting Species Survival in the Wild*. Island Press, Washington, pp. 40–67.

CPC (Center for Plant Conservation). 1991. Genetic sampling guidelines for conservation collections of endangered plants. In: Falk D.A., Holsingers K.E. (eds) *Genetics and Conservations of Rare Plants*. Oxford University Press, New York, pp. 225–238.

Goel, A.K. 1992. Observations on habitats of some rare and threatened plants in Bhillangana Valley of Tehri Garhwal. *J. Econ. Taxon. Bot.* 16(1): 193-198.

Guerrant E.O., Fiedler P.L., Havens K., Maunder M. 2004. Revised genetic sampling guidelines for conservation collections of rare and endangered plants. In: Guerrant E.O., Havens K., Maunder M. (eds) Ex Situ *Plant Conservation Supporting Species Survival in the Wild*. Island Press, Washington, pp. 419–442.

Mamgain S. K., 1999. Phenological observation and conservation of *Sophora mollis* Royle (Papilionaceae) an endangered multi-purpose legume of North-West Himalaya. *Taiwania*, 44(1):137-144.

Prakash Anand and Singh, K.K. 2001. Observation on some threatened plants and their conservation in Rajaji National Park, Uttaranchal, India. *J. Econ.Tax. Bot.* 25 (2): 363-366.

Prakash Anand, KK Rawat and PC Varma, 2009. Indopiptadenia oudhensis (Brandis) Brenan: monotypic, endemic and highly endangered taxa needs conservation in Uttar Pradesh. Biodiv News, U.P. State Biodiversity Board, Vol 1, Issue 1, p.g 2-5.

Saha, D., Ved, D., Haridasan, K. and Ravikumar, K. 2015. *Nepenthes khasiana*. The IUCN Red List of Threatened Species 2015: e.T48992883A49009685. http://dx.doi.org/10.2305/IUCN.UK.2015-2.RLTS.T48992883A49009685.en

Srivastava S. K. 1985. Jasminum parkeri Dunn – an endangered, endemic taxon from Himachal Pradesh. *J. Econ. Taxon. Bot.* 7(3): 709-710.

Walter, K.S. and H.J. Gillet. (eds.). 1998. 1997 IUCN Red List of Threatened Plants. The World Conservation Monitoring Centre IUCN-The World Conservation Union, Gland, Switzerland and Cambridge, UK. 1xvi+862 pp.

Chapter 8

Wild Medicinal Plant Species in Botanic Garden of Indian Republic, Noida

Sheo Kumar[1] *and Damini Sharma*[2]

[1]*Botanic Garden of Indian Republic, Botanical Survey of India, Govt. of India Ministry of Environment, Forest and Climate Change, Capt. Vijyant Thapar Marg, Sector 38A, NOIDA – 201 303, U.P.*
[2]*University School of Environment Management, Guru Gobind Singh Indraprastha University, Dwarka, New Delhi – 110 075*
E-mai: drsheo_kumar@rediffmail.com

Abstract

As per Article 13 of the Convention of Biological Diversity (CBD) under Training, Public Education and Awareness programme of Botanic Garden of Indian Republic (BGIR), NOIDA inventorisation of wild plants, preparation of Herbarium, identification along medicinal properties present in them were worked out during June and July, 2015 as per recommended methods and techniques. Altogether 36 wild plant species, among 30 species recorded as Dicot [number-wise 6 species of Fabaceae, 5 species of Asteraceae, 3 species each of Amaranthaceae, Euphorbiaceae and Malvaceae, 2 species of Solanaceae and 1 species each of Asclepiadaceae, Cannabaceae, Cucurbitaceae, Meliaceae, Moraceae, Rhamnaceae, Verbenaceae and Zygophyllaceae] and 6 species recorded as Monocot [number-wise 5 species of Poaceae and 1 species of Cyperaceae] were inventorised within the premises of BGIR randomly from five different stations and three sites each. Based on field observations and publications of identified specimens, their medicinal properties along with their distribution, plant type and ecological characteristics ascertained for carrying out further research and conservation for future generation.

***Keywords**: CBD, Training, Education, Awareness and Wild medicinal plants.*

Introduction

India is one of richest reservoirs of biological diversity in the world. It possess a great variety of ethno-medicinal plant species and is ranked 6th (sixth) among 12 mega diversity countries of the world. Medicinal plants are such plants which are having certain potent constituents in the form of metabolites *i.e.* organic compounds including terpenoids, special nitrogen metabolite (like non-protein amino acids, amines, cyanogenic glycosides, glucosinolates and alkaloids) tannins, resins, taraxerol, phenolics, volatiles oils and others which are not utilised by plant itself for their development, vary in plants of different Family differently in specific tissues, cells or developmental stages during whole tenure of plant development. These metabolites and or other deposits on their application individually and or in combination cure illness/disease. They range from those used in production of mainstream pharmaceutical products to plants used in herbal medicine preparations. In all cases, wild plant species play a major role in forming foundation of healthcare practices maximum in Asia are traditional Chinese medicine, Ayurveda, Siddha, Unani and Tibetan medicines. Over 3,500 wild species are harvested by people who are skilled in plant identification to use them as folk-medicines to cure ailments in man and their domesticated animals. This practice, use of wild plants as medicines is pre-dates as written in human history. Hence, it has been recognized as an effective way to discover future medicines. As per World Health Organization (WHO) about 70 per cent to 80 per cent of the world population in the developing countries depends on herbal drugs under primary health care system. **India is major hub of wild-plant medicine industry in Asia, as a result, key species have been declined due to over-exploitation.** Thus, 93 per cent of wild medicinal plants used for making herbal medicines in the country are endangered and government is trying to relocate them from their usual habitat to protect them. Besides, in general, other floral forms are also under threat and need their conservation and restoration in respective habitats.

Keeping these into considerations a 'National Botanic Garden' (now known as 'Botanic Garden of Indian Republic') was established in NOIDA during 1997 by the Ministry of Environment, Forest and Climate Change through Botanical Survey of India. The garden is dedicated for conservation of endemic and threatened plants of different phyto-geographical regions of the country under *ex-situ* conservation along with other components like research, training, environmental education/ awareness and recreation.

Before establishment of this garden, land area was under cultivation of cereals [*Pennisetum glaucum* (L.) R.Br. (Bajra), *Sorghum bicolor* (L.) Moench (Jowar), *Zea mays* L. (Maize), *Avena sativa* L. (Oat), *Oryza sativa* L. (Paddy) and *Triticum* aestivum L. (Wheat); pulses [*Cicer arietinum* L. (Gram) and *Pisum sativum* L. (Pea)]; mustard (*Brassica nigra* L.); vegetables [*Solanum melongena* L. (Brinjal), *Brassica oleracea* var. *capitata* (Cabbage), *Brassica oleracea* var. *botrytis* (Cauliflower), *Cucurbita moschata* Duchesne ex Poir. (Cucurbits), *Lagenaria siceraria* (Molina) Standl. (Lauki), *Cucumis sativus* L. (Kheera), *Abelmoschus esculentus* L. (Lady's Finger), *Allium cepa* L. (Onion), *Solanum tuberosum* L. (Potato) and *Raphanus sativus* L. (Radish)]; fodder [*Trifolium alexandrinum* L. (Berseem)]; natural vegetation [*Dalbergia sissoo* Roxb.

(Shisham), *Cassia fistula* L. (Cassia), *Azadirachta indica* A.Juss. (Neem) and *Morus alba* L. (Morus)] and planted plants for greening the areas were *Parkinsonia aculeate* L. (Parkinsonia), *Thevetia peruviana* (Yellow Kaner/Oleander) and *Eucalyptus tereticornis* Sm. (Eucalyptus), *etc.* was sparse. However, so far, more than 10,500 individuals of 876 plant species (comprising Pteridophytes (25), Gymnosperms (11) and Angiosperms (840) including 37 species are listed in Red Data Book and more than 250 medicinal plants brought from 23 States of the country are conserved under *ex-situ* Conservation in Arboretum (*Green Belt, Economic Plants Garden and Fruits Garden*), Nurseries, Thematic Gardens (*Map of India, Cactus and Succulents Garden, Medicinal Garden, Landscaped Garden*), Horticulture Gardens (*Orchids, Ferns, Cycades, Rosaries, Bougainvillea's and Seasonal Gardens*) and Water bodies (BGIR, 2015). Besides, due to succession, various wild species have naturally occurred in the garden and are unexplored for their beneficial properties present in them. Keeping these into consideration, present scientific work was carried out for finding scientific information for addition in existing scientific knowledge for future reference as well as to meet the requirement of Article 7, 9 and 13 of CBD.

Materials and Methods

The study areas were randomly selected at five different locations within premises of Botanic Garden of Indian Republic, situated at 28°33′27″ N. lat. to 28°34′00″ N. lat. and 77°19′16″ E. long. to 77°20′13″ E. long. and at altitude above 197 to 202 MSL depicted in Figure 8.1. Meteorological characteristics recorded at nearby Safdarjung Weather Station locate at 28°58′ N. lat. and 77°2′ E. long. above 211 MSL was downloaded from departmental website and referred.

Figure 8.1: Panoramic Satellite View of Botanic Garden of Indian Republic, BSI, MOEF and CC, Govt. of India, NOIDA and Sampling Stations.

Field work was undertaken by walk-over survey and plants in wild conditions were visually inspected for distribution, recorded general plant habit like herb, shrub, tree, annual, perennial, erect, scandent followed by preliminary identification which was undertaken by studying macroscopic morphological and reproductive characteristics like root, stem, leaf, inflorescence, flower, fruit, seed along with their size and height as well as studying organoleptic characteristics or sensory characteristics *e.g.*, colour, texture (touching, the softness or hardness of the plant parts examined), taste and mouth feel (specific plant only) and odor (aromatic, fruity, musty, mouldy, rancid, *etc.*) for confirming plant identity as described by Applequist (2006). For further identification, the following steps of described methods were also applied:

1. Collection of Plant Samples

Preliminarily, inventory of wild plant species occurred within a radius of 15 m at each selected station and three sites was undertaken. Whole plant specimen (except trees) including roots, branches, leaves, inflorescence, flowers, *etc.* were collected in polyethylene bags and brought to the laboratory, washed with raw water followed by preserving them in mixture of Formaldehyde and Absolute Alcohol. Excess solvent was taken out by drying them under shade for short period. In the field, each specimen assigned a collection number, entry of information of the locality of collection, date, collected by, *etc.* for each specimen entered in field notebook (Maden, 2004). The specific collection number assigned to each specimen also written on folded paper containing specimen on completion of Herbarium.

2. Preparation of Herbarium

Processed specimens were pressed in Herbarium press enough tight under blotting papers as per Maden (2004) for shortest time period as long as with changing blotting papers frequently to get absorbed moisture content of specimens and to prevent any shrinkage as well as wrinkling in plant material without damaging the same to get properly dried specimen for easy mounting. Fully dried specimens were then mounted on herbarium paper (an acid-free cardstock) for their long-term storage. Collection number and label with recorded information pasted at specific location in each specimen and kept in Herbarium folder for final identification and preservation.

3. Identification of Specimen and Beneficial Properties

The specimens were finally identified consulting relevant literatures, Monographs and other referred books (Choudhury and Pandey, 2007; Vardhana, 2007) and medicinal properties present in them were further ascertained considering field observations and consulting referred literatures (Ambasta, 2000; Chopra *et al.*, 2005; Dhiman, 2003, 2005; Kala, 2003; Kaushik and Dhiman, 2000; Kumar *et al.*, 2006; Sinha and Jain, 2001; Vishwanathan *et al.*, 2006). Besides, medicinal properties of individual species were recorded as per information available and compiled for specific species and cited after the text.

Results and Discussion

During this study, meteorological data of each day for two months *i.e.* June and July, 2015 were collected and compiled for knowing minimum, maximum and average value and depicted in Table 8.1. The number of identified wild plant species belonging to different Family were worked out and presented in Table 8.2 and Figure 8.2.

Table 8.1: Meteorological Data Recorded at Safdarjung Weather Station, New Delhi

Parameters	*June, 2015*			*July, 2015*		
	Temperature (°C)		*Rainfall (mm)*	*Temperature (°C)*		*Rainfall (mm)*
	Min	*Max.*		*Min*	*Max.*	
Min.	21	31	0	23	27	0
Max.	32	43	22	31	40	133
Average	26.4	38.6	2.13	26.0	34.7	7.4
SD (±)	2.66	3.02	5.10	1.89	3.14	23.87
No. of Rain days			9			12

Table 8.2: Number of Identified Species Belongs to different Families

Sl.No.	*Species belong to Different Family*	*No. of Species*
Dicot		
1	Fabaceae	6
2	Asteraceae	5
3	Amaranthaceae	3
4	Euphorbiaceae	3
5	Malvaceae	3
6	Solanaceae	2
7	Asclepiadaceae	1
8	Cannabaceae	1
9	Cucurbitaceae	1
10	Meliaceae	1
11	Moraceae	1
12	Rhamnaceae	1
13	Verbenaceae	1
14	Zygophyllaceae	1
Monocot		
	Poaceae	5
	Cyperaceae	1
Total species		**36**

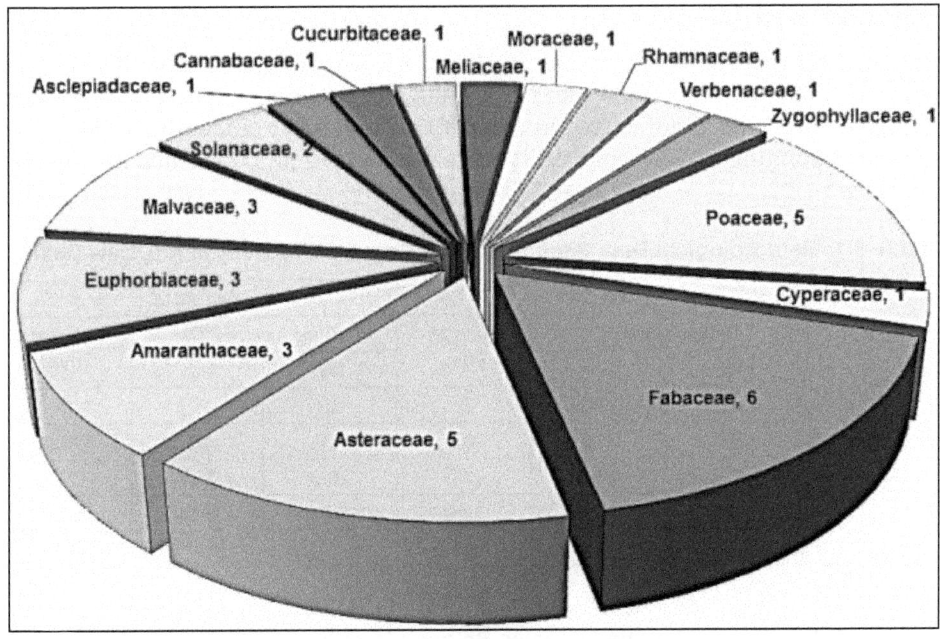

Figure 8.2: Number of Identified Species Belong to different Family of Monocot and Dicot.

During the study, altogether 36 species were identified by studying Herbarium of collected specimens. Among 30 species were recorded as Dicot [6 species of Fabaceae, 5 species of Asteraceae, 3 species each of Amaranthaceae, Euphorbiaceae and Malvaceae, 2 species of Solanaceae and 1 species each of Asclepiadaceae, Cannabaceae, Cucurbitaceae, Meliaceae, Moraceae, Rhamnaceae, Verbenaceae and Zygophyllaceae] and 6 species recorded as Monocot [5 species of Poaceae and 1 species of Cyperaceae] presented in Table 8.2 and Figure 8.2. Out of identified 36 species, 10 species namely *Abutilon indicum* (Link) Sweet, *Achyranthes aspera* L., *Azadirachta indica* A.Juss., *Calotropis procera* (Aiton) W.T.Aiton, *Cannabis sativa* L., *Clitoria ternatea* L., *Dalbergia sissoo* Roxb., *Eclipta prostrata* (L.) L., *Ficus religiosa* L. and *Sida acuta* Burm.f. have already been conserved under *ex-situ* conservation in 'Ayur Vatika' (Medicinal Section) and other areas of the garden and thus Article 9 of CBD also fulfilled. Once, identity of plant species confirmed, they were further studied for medicinal properties present in them through published literatures and described individually are as follows:

Dicots

Abutilon indicum (Link) Sweet (Family Malvaceae) Coll.No.: 2004

Vernacular Name: Kanghi

Status: Common weed, considered as invasive; conserved in Medicinal Section.

Medicinal Properties: Traditional medicine; in leprosy, ulcers, headaches, gonorrhea and bladder infection; as demulcent, aphrodisiac, piles, laxative, diuretic, sedative, astringent, expectorant, tonic, anti-inflammatory, anti-helminthic and analgesic (Rajakaruna *et al.*, 2002).

Achyranthes aspera L. (Family Amaranthaceae) Coll.No.: 2005

Vernacular Name: Latjira

Status: Common weed; conserved in Medicinal Section.

Medicinal Properties: In toothache, cough, obstetrics and gynecology including abortion, induction of labor, cessation of postpartum bleeding and ease symptoms of malaria (Bussmann *et al.*, 2006).

Alhagi maurorum Medik. (Family Fabaceae) Coll.No.: 2006

Vernacular Name: Camelthorn bush

Status: Indigenous, noxious weed.

Medicinal Properties: Folk medicine; in glandular tumors, rheumatism, hemorrhoids, nasal polyps, ailments related to bile ducts; as gastro-protective, diaphoretic, diuretic, expectorant, laxative, anti-diarrhoeal, anti-septic and sweetener (USDA, 2012).

Amaranthus viridis L. (Family Amaranthaceae) Coll.No.: 2011

Vernacular Name: Jungali chaulayi

Status: Common weed.

Medicinal Properties: Traditional Ayurvedic medicine; in emollient, leucorrhoea and leprosy (Nair, 2003).

Azadirachta indica A.Juss. (Family Meliaceae) Coll.No.: 2017

Vernacular Name: Neem

Status: Weed; noted for its drought resistance; conserved in various sections and along road of the garden.

Medicinal Properties: Siddha, Ayurvedic and Unani medicines; in skin diseases like eczema, psoriasis, *etc.*, healthy hair, liver function, blood purification and balance blood sugar levels, *etc.*; as anti-helminthic, anti-fungal, anti-diabetic, anti-bacterial, anti-viral, anti-septic, anti-cancer contraceptive and sedative; long-term use may harm kidneys or liver and cause miscarriages, infertility and low blood sugar; oil toxic to infant and may lead to death (Biswas *et al.*, 2002).

Calotropis procera (Aiton) W.T.Aiton (Family Asclepiadaceae) Coll. No.: 2002

Vernacular Name: Aak

Status: Weed; conserved in Medicinal Section.

Medicinal Properties: In boils, eczema, leprosy, cutaneous infections, cough and colds, asthma, piles, inflammation, rheumatism, syphilis, elephantiasis, malarial and low hectic fevers and pyrexias; as abortifacient, anti-inflammatory, anti-microbial, anti-malarial, hepato-protective agents, analgesic and cytostatic.

Cannabis sativa L. (Family Cannabaceae) Coll.No.: 2006

Vernacular Name: Ganja, Hemp

Status: Weed; conserved in Medicinal Section.

Medicinal Properties: In nausea, vomiting, pain, neurological problems, HIV/AIDS; as analgesic, hallucinogenic, hypnotic, sedative, anti-convulsant and anti-inflammatory agent (Wang *et al.*, 2014 and Whiting *et al.*, 2015).

Clitoria ternatea L. (Family Fabaceae) Coll.No.: 2027

Vernacular Name: Aparajita

Status: Weed; grows well in moist and neutral soil; conserved in Medicinal Section.

Medicinal Properties: Traditional medicine; in headache, leukoderma, sexual ailments like infertility, gonorrhea and control of menstrual discharge; as memory enhancer, nootropic, anti-stress, anxiolytic, anti-depressant, anti-convulsant, tranquilizing, aphrodisiac and sedative agent (Mukherjee *et al.*, 2008).

Corchorus olitorius L. (Family Malvaceae) Coll.No.: 2022

Vernacular Name: Nalta or Tossa jute

Status: Weed; grows well in all kinds of soil and pH, prefer shade and moist soil.

Medicinal Properties: In chronic cystitis, gonorrhea, dysurial, restoration of appetite and strength and improvement of cardiac insufficiencies; as purgative, demulcent, diuretic, febrifuge and tonic.

Croton bonplandianum Baill. (Family Euphorbiaceae) Coll.No.: 2013

Vernacular Name: Ban tulsi

Status: Weed; invasive.

Medicinal Properties: In skin diseases (like ringworm, swellings in body), liver, cuts and wounds, cholera, jaundice, acute constipation, abdominal dropsy, internal abscesses, lowers blood pressure and alleviates spasms (Singh *et al.*, 2014).

Dalbergia sissoo Roxb. (Family Fabaceae) Coll.No.: 2029

Vernacular Name: Shisham, Sisu

Status: Prefers soil sandy, gravel to rich alluvium of river banks and slightly saline condition; conserved in various sections and along road of the garden.

Medicinal Properties: In skin diseases like boils/eruptions, leprosy, reducing plaque, gingival inflammation, dysentery and gonorrhea; as astringent (Bhambal *et al.*, 2011).

Datura inoxia Mill. (Family Solanaceae) Coll.No.: 2018

Vernacular Name: Dhatura

Status: Weed

Medicinal Properties: In asthma, disinfecting wounds by poultices, diarrhoea, fever and treatment of broken bones, swollen joints as well as pain; as anodyne, aphrodisiac, sedative and anti-spasmodic.

***Digera muricata* D. (L.) Mart. (Family Amaranthaceae) Coll. No.: 2020**

Vernacular Name: Lata mouri

Status: Endemic; weed.

Medicinal Properties: In biliousness and urinary discharges; as laxative, cooling and astringent to bowels (Yusuf *et al.*, 2009).

***Eclipta prostrata* (L.) L. (Family Asteraceae) Coll. No.: 2023**

Vernacular Name: Bhringaraj

Status: Endemic; weed; conserved in Medicinal Section.

Medicinal Properties: In preventing hair loss, miscarriage and post-delivery pain (Puri, 2003).

***Erigeron bonariensis* L. (Family Asteraceae) Coll.No.: 2019**

Vernacular Name: Flax-leaf Fleabane

Status: Weed; alien.

Medicinal Properties: In stopping bleeding, diarrhoea, dysentery, diabetes, scalding urine and haemorrhage of bowels, uterus and wounds, throat infection, cough and fever (Tilley, 2012).

***Euphorbia hirta* L. (Family Euphorbiaceae) Coll. No.: 2010**

Vernacular Name: Dudhi, Asthma weed

Status: Weed

Medicinal Properties: Traditional medicine; in female disorders, respiratory ailments (cough, coryza, bronchitis and asthma), conjunctivitis, worm infestations in children, amoebic dysentery, jaundice, pimples, gonorrhea, digestive problems and tumors (CSIR, 2005 and Kumar *et al.*, 2010).

***Euphorbia thymifolia* L. (Family Euphorbiaceae) Coll.No.: 2012**

Vernacular Name: Chhoti Duddhi

Status: Weed

Medicinal Properties: In piles, ringworm, skin diseases, aggravated cough, amenorrhoea, remedy for snake bites; as diuretic, laxative, detumescent, anti-malarial, anti-rash, anti-viral, anti-oxidant; anti-carbuncle, anti-hemorrhoidal, sedative, blood purifier, hemostatic, stimulant, aphrodisiac, detoxificant, expectorant, astringent in diarrhoea and dysentery and possesses age sustaining properties (Mali and Panchal, 2013).

Ficus religiosa L. (Family Moraceae) Coll. No.: 2035

Vernacular Name: Peepal

Status: Endemic; thrive in hot and humid weather; prefers full sunlight and pH 7 or less; grows well in most soil types among loam is best; conserved in premises of the garden.

Medicinal Properties: Traditional medicine; in asthma, diabetes, diarrhoea, epilepsy, gastric problems, inflammatory disorders, infectious, sexual disorders, *etc*. (Singh *et al.*, 2011).

***Lantana camara* var. *aculeate* (Family Verbenaceae) Coll. No.: 2003**

Vernacular Name: Raimuniya; Wild sage

Status: Noxious weed; invasive.

Medicinal Properties: Traditional medicines; in ailments like skin itches, leprosy, rabies, chicken pox, measles, headache, asthma, ulcers and cancer; as anti-microbial, fungicidal and insecticidal properties (Barreto *et al.*, 2010 and Sathish *et al.*, 2011).

***Launaea asplenifolia* (Wild) Hook.f. (Family Asteraceae) Coll. No.: 2025**

Vernacular Name: Balrajkonda, Tik chana, Titlia

Status: Weed

Medicinal Properties: In bilious complaints, fever, cough, asthma demulcent; as stimulant, enhance stamina for day long hard works whereas due to its use persistent useless libido leads to sterility (RPMC, 2014).

***Lysiloma latisiliquum* (L.) Benth. (Family Fabaceae) Coll. No.: 2032**

Vernacular Name: Wild tamarind, Subabul

Status: Weed

Medicinal Properties: So far medicinal properties not known.

***Mukia maderaspatana* (L.) M.Roem. (Family Cucurbitaceae) Coll. No.: 2009**

Vernacular Name: Wild Cucurbit, Melon-gubat, Agumaki

Status: Weed

Medicinal Properties: Traditional medicines like Folkloric, Siddha and Ayurvedic systems for over three centuries in treatment of vertigo, biliousness, jaundice, cough, cold and flue; as aperient, diuretic, stomachic, antipyretic, antiflatulent, antiasthmatic, antitussive, antihistaminic, antibronchitic, antiinflammatory, antimicrobial, hepatoprotective and expectorant (Petrus *et al.*, 2011).

***Nicotiana plumbaginifolia* Viv. (Family Solanaceae) Coll. No.: 2033**

Vernacular Name: Ban tambaku

Status: Weed; optimal soil texture in acceptable soil pH.

Medicinal Properties: In skin diseases, rheumatic pain, swelling, nausea and travel sickness, veterinary medicine; as germicide, antispasmodic, antibacterial, diuretic and expectorant (Singh *et al.*, 2010).

Parthenium hysterophorus L. (Family Asteraceae) Coll. No.: 2014

Vernacular Name: Ramphool, Congress Grass or Gajar Ghans

Status: Noxious weed

Medicinal Properties: In skin inflammation, rheumatic pain, diarrhoea, urinary tract infections, dysentery, malaria and neuralgia (Patel, 2011).

Pluchea lanceolata (DC.) Oliv. and Hiern (Family Asteraceae) Coll. No.: 2008

Vernacular Name: Indian Camphorweed, Rasna

Status: Common weed; grows well in Sandy areas.

Medicinal Properties: Traditional medicine; in constipation, respiratory diseases, swellings of joint in arthritis, rheumatism, neurological diseases, inflammations, bronchitis, psoriasis, cough, thermogenic, allaying pain caused by sting of scorpions and piles; as aperients, laxative, analgesic, antipyretic, nervine tonic and cooling agent during summer (Srivastava and Shanker, 2012).

Prosopis juliflora (Sw.) DC. (Family Fabaceae) Coll. No.: 2031

Vernacular Name: Vilayati kikar or babul

Ecology and *Status*: Weed; invasive.

Medicinal Properties: In common problems of eyes, cough, laryngitis, dermatological ailments like sores, wounds, burns, sunburn, chapped fingers, lips and intestinal problems including stomach inflammation, diarrhoea, system cleansing or settling intestine, painful gums, bladder infection, hemorrhoids, measles or fever, headaches, red ant and animal stings; as an antacid to treat digestive problems, anti-bacterial, anti-septic, soothing, astringent, purgatives, eyewashes to treat pink eye, infection and irritation.

Senna occidentalis (L.) Link (Family Fabaceae) Coll. No.: 2001

Vernacular Name: Coffee weed, Chakunda

Status: Weed; invasive.

Medicinal Properties: In cough, asthma, bronchitis, fever, convulsion, edema, urinary tract disorders, detoxifying liver, gonorrhea, internal bacterial and fungal disorders, viral infections, intestinal parasites, enhancement of immunity and promotes perspiration, reduction of blood pressure and spasms; as purgative, laxative, expectorant, anti-inflammatory, anti-malarial, hepato-protective, cardio-tonic, analgesic, febrifuge, vermifuge and adulterant of coffee.

Sida acuta Burm.f. (Family Malvaceae) Coll.No.: 2030

Vernacular Name: Broom grass, Pilla valatti chedi

Status: Common weed, conserved in Medicinal Section.

Medicinal Properties: Traditional medicine; in microbial infections, dysentery, vomiting, gastric disorders, bronchitis, asthma, fever, pains, ulcers, skin and venereal diseases, dandruffs and snake bites; as febrifuge, abortifacient, diuretic, anti-worm and antibacterial (Oboh *et al.*, 2007).

***Tribulus terrestris* L. (Family Zygophyllaceae) Coll. No.: 2028**

Vernacular Name: Gokshura, Bindii

Status: Noxious weed; invasive.

Medicinal Properties: Ayurveda, Unani and Chinese traditional medicine system; in back pain, inflammation, sciatica, respiratory disorders, dry cough, dysuria, heart diseases, diabetes, dyspnoea, blood disorders, haemorrhoids, debility, burning sensation, digestive system, improvement of renal discomfort including kidney, bladder, urinary tract/calculi, urogenital for overall male sexual performance, building of all muscle and tissues especially shukra dhatu (reproductive tissue) but it is not considered as a particular vajikarana (sexual functions); as diuretic, nerve tonic, aphrodisiac and rejuvenative (i-AIM, 2011 and Kevalia and Patel, 2011).

***Ziziphus nummularia* (Burm.f.) Wight and Arn. (Family Rhamnaceae) Coll. No.:2015**

Vernacular Name: Jharber

Status: Endemic; Thar Desert of Western India; weed; found usually in agricultural field.

Medicinal Propertie: In pyorrhea, eczema, scabies and other skin diseases; as astringent in bilious affliction and antimicrobial (Gautam *et al.*, 2011).

Monocot

***Aristida adscensionis* L. (Family Poaceae) Coll. No.: 2027**

Vernacular Name: Six weeks threeawn

Status: Weed; invasive; tolerant to drought; grow on all soils.

Medicinal Properties: Ethno-medicines; in itching and ringworm; as lactation (Heuzé *et al.*, 2015).

***Cynodon dactylon* (L.) Pers. (Family Poaceae) Coll. No.: 2021**

Vernacular Name: Doob, Star grass

Status: Weed; invasive; grow on saline soils and under saline watering.

Medicinal Properties: As per Ayurveda, Unani, Homoeopathic and traditional pharmacopoeia, this sacred grass is used in treatment of bleeding of gums, toothache, remove bad odour from the mouth (halitosis), cold and cough, asthma, bronchitis, stomach ailments like acidity, ulcer, colitis, constipation, indigestion, stomach infections and pain, diarrhea, optimizing immune system, piles, urticaria, injuries, eye and nose problems, wounds, skin rashes, itching, eczema, leprosy, scabies, leucoderma, mental debility, diabetes, epilepsy, nervous system, weight loss, stress, inflammations, urinary tract infections, vaginal problems, menstrual problems, gynecological problems, syphilis, tumors and breast milk production; as antiviral, antimicrobial, antiseptic, anti-inflammatory, anthelmintic, antipyretic, anticancer, alexiteric, appetizer, laxative, brain and heart tonic, aphrodisiac, alexipharmic, emetic, emmenagogue, expectorant and carminative (GC, 2014 and Oudhia, 2001).

Cyperus rotundus L. (Family Cyperaceae) Coll. No.: 2034

Vernacular Name: Motha, Nut grass

Status: Weed; prefers dry conditions even tolerate moist soils, often grows in wastelands/in crop fields.

Medicinal Properties: In fever, digestive systems disorder, dysmenorrhea and other maladies; as anti-microbial, anti-malarial, anti-oxidant and anti-diabetic (Jagtap *et al.*, 2004 and Patel *et al.*, 2010).

Panicum capillare L. (Family Poaceae) Collection No.: 2024

Vernacular Name: Witch grass

Status: Weed; invasive; grows well in all kinds of soil, prefers well-drained moist soil and pH, perish under shade.

Medicinal Properties: As emetic and reducing aid when dieting.

Pennisetum ciliare (L.) Link (Family Poaceae) Coll. No.: 2016

Vernacular Name: Anjan Grass

Status: Weed

Medicinal Properties: Folk medicine; in kidney pain, tumors, sores and wounds; as pain reliever, diuretic, emollient and a lactagogue.

Setaria verticillata (L.) P.Beauv. (Family Poaceae) Coll. No.: 2036

Vernacular Name: Barchittas, Latkaunya, Laptuna

Status: Noxious weed; invasive; grows well in all kinds of low and fertile soil within pH ranging from 6.1 to 8.0.

Medicinal Properties: Folk medicine; in treating joint pains, diarrhoea and scorpion bite.

Conclusion

Based on above observations, out of identified and inventorised 36 species except *Lysiloma latisiliquum* (L.) Benth., all 35 species are possessing medicinal properties. Among all, sacred grass *i.e. Cyanodon dactylon* (L.) Pers. has been enlisted to cure maximum illness/diseases followed by *Tribulus terrestris* L., *Prosopis juliflora* (Sw.) DC., *Senna occidentalis* (L.) Link., *Azadirachta indica* A. Juss. and other species. Though, these species are wild and invasive/alian in nature, however, as they are possessing potent medicinal properties in them, 10 species have already been conserved in the garden under *ex-situ* conservation for future generation.

Acknowledgement

Authors are thankful to the Director, Botanical Survey of India, Kolkata for constant inspiration and encouragements and to the Ministry of Environment, Forest and Climate Change for providing logistic support and opportunity to carry out this work.

References

Ambasta S.P. 2000. The Useful Plants of India. National Institute of Science Communication, New Delhi.

Applequist W. 2006. The Identification of Medicinal Plants: A Handbook of the Morphology of Botanicals in Commerce. American Botanical Council, Austin.

Barreto F.S., Sousa E.O., Campos A.R., Costa J.G.M. and Rodrigues F.F.G. 2010. Antibacterial Activity of *Lantana camara* L. and *Lantana montevidensis* Brig extracts from Cariri-Ceará, *Journal of Young Pharmacists* 2(1): 42–44.

BGIR 2015. Brochure on Memorandum of Understanding between MOEF and CC and NOIDA. June, 5th, 2015. Botanic Garden of Indian Republic, Govt. of India, NOIDA.

Bhambal A., Kothari S., Saxena S. and Jain M. 2011. Comparative effect of neem stick and toothbrush on plaque removal and gingival health – A clinical trial. *Journal of Advanced Oral Research 2(3): 51–56.*

Biswas K., Chattopadhyay I., Banerjee R.K. and Bandyopadhyay U. 2002. Biological activities and medicinal properties of Neem (*Azadirachta indica*). *Current Science* 82(11): 1336–1345.

Bussmann R.W., Gilbreath G.G., Solio J., Lutura M., Lutuluo R., Kunguru K., Wood N. and Mathenge S.G. 2006. Plant use of the Maasai of Sekenani Valley, Maasai Mara, Kenya. *Journal of Ethnobiology and Ethnomedicine* 2: 22.

Chopra R.N., Chopra I.C. and Verma B.S. 2005. Supplement to Glossary of Indian Medicinal Plants. National Institute of Science Communication, New Delhi.

Chowdhery H.J. and Pandey D.S. 2007. Plants of Indian Botanic Garden. Botanical Survey of India, Howrah.

CSIR 2005. The Wealth of India (Raw Material). Council of Industrial and Scientific Research, New Delhi.

Dhiman A.K. 2003. Sacred Plants and their Medicinal Uses. Daya Publishing House, New Delhi.

Dhiman A.K. 2005. Wild Medicinal Plants of India. Bishen Singh and Mahendra Pal Singh, Dehradun.

Gautam S., Jain A.K. and Kumar A. 2011. Potential antimicrobial activity of *Zizyphus nummularia* against medically important pathogenic microorganisms. *Asian Journal of Traditional Medicines* 6(6): 267–271.

GC 2014. Amazing Health Benefits and Medicinal Uses of Bermuda Grass (*Cynodon dactylon*). Gyanunlimited.com-A Hub of Alternative Medicine and Holistic Health, AYUSH.

Heuzé V., Tran G., Lebas F. and Maxin G. 2015. *Common needle grass (Aristida adscensionis)*. Feedipedia, a programme by INRA, CIRAD, AFZ and FAO.

i-AIM 2011. Goksura (*Tribulus terrestris*). DST-DPPRP funded National R&D Facility for Rasayana products of Indian Systems of Medicine. Institute of Ayurveda and Integrative Medicine, Bangalore.

Jagtap A.G., Shirke S.S. and Phadke A.S. 2004. Effect of polyherbal formulation on experimental models of inflammatory bowel diseases. *Journal of Ethnopharmacology* 90(2-3): 195–204.

Kala C.P. 2003. Medicinal Plants of Indian Trans-Himalaya. Bishen Singh and Mahendra Pal Singh, Dehradun.

Kaushik P. and Dhiman A.K. 2000. Medicinal Plants and Raw Drugs of India. Bishen Singh and Mahendra Pal Singh, Dehradun.

Kevalia J. and Patel B. 2011. Identification of fruits of *Tribulus terrestris* L. and *Pedalium murex* L.: A pharmacognostical approach. *Ayu* 32(4): 550–553.

Kumar J., Rout, S.D. and Das M.K. 2006. The Medicinal Plants of Hatikote Forests of District Mayurbhang, Orissa. *Need for Conservation* 132(1): 43-53.

Kumar S., Malhotra R. and Kumar D. 2010. *Euphorbia hirta:* Its chemistry, traditional and medicinal uses and pharmacological activities. *Pharmacognosy Reviews* 4(7): 58–61.

Maden K. 2004. Plant Collection and Herbarium Techniques. *Our Nature* 2: 53–57.

Mali P.Y. and Panchal S.S. 2013. A review on phyto-pharmacological potentials of *Euphorbia thymifolia* L. *Ancient Science of Life* 32(3): 165–172.

Mukherjee P.K., Kumar V., Kumar N.S. and Heinrich M. 2008. The Ayurvedic medicine *Clitoria ternatea* from traditional use to scientific assessment. *Journal of Ethnopharmacology* 120(3): 291-301.

Nair R.V. 2003. Controversial Drug Plants. Universities Press, Hyderabad.

Oboh *I.e.*, Akerele J.O. and Obasuyi O. 2007. Antimicrobial activity of the ethanol extract of the aerial parts of *Sida acuta* Burm.f. (Malvaceae). T*ropical Journal of Pharmaceutical Research,* 6(4): 809–813.

Oudhia P. 2001. Doob (*Cynodon dactylon*): Traditional Medicinal Uses in India. NewCROP Center, Purdue University. Society for Parthenium Management (SOPAM), Raipur.

Patel M.V., Patel K.B. and Gupta S.N. 2010. Effects of Ayurvedic treatment on forty-three patients of ulcerative colitis. *Ayu* 31(4): 478-481.

Patel S. 2011. Harmful and beneficial aspects of *Parthenium hysterophorus:* an update. *3 Biotech* 1(1): 1–9.

Petrus A.J.A., Bhuvaneshwari N. and Alain J.A.L. 2011. Antioxidative constitution of *Mukia maderaspatana* (L.) M.Roem. leaves. *Indian Journal of Natural Products and Resources* 2(1): 34–43.

Puri H.S. 2003. Rasayana: Ayurvedic Herbs for Longevity and Rejuvenation. Taylor and Francis, London, pp. 80 – 85.

Rajakaruna N., Harris C.S. and Towers G.H.N. 2002. Antimicrobial Activity of Plants Collected from Serpentine Outcrops in Sri Lanka. *Pharmaceutical Biology*, 40(3): 235–244.

RPMC 2014. *Launaea asplenifolia* Hook.f. Raja Peary Mohan College, Hooghly.

Sathish R., Vyawahare B. and Natarajan K. 2011. Antiulcerogenic activity of *Lantana camara* leaves on gastric and duodenal ulcers in experimental rats. *Journal of Ethnopharmacology* 134(1): 195–197.

Singh D., Singh B. and Goel R.K. 2011. Traditional uses, phytochemistry and pharmacology of *Ficus religiosa*: a review. *Journal of Ethnopharmacology* 134(3): 565–583.

Singh K.P., Daboriya V., Kumar S. and Singh S. 2010. Antibacterial activity and phytochemical investigations on *Nicotiana plumbaginifolia* Viv. (Wild tobacco). *Romanian Journal of Biology - Plant Biology* 55(2): 135–142.

Singh N.K., Ghosh A., Laloo D. and Singh V.P. 2014. Pharmacognostical and physico-chemical evaluation of *Croton bonplandianum. International Journal of Pharmacy and Pharmaceutical Sciences* 6(3): 286 290.

Sinha R.K. and Jain S.K. 2001. Ethnobotany: The Renaissance of Traditional Herbal Medicine. INA Shree Publishers, Jodhpur.

Srivastava P. and Shanker K. 2012. *Pulchea lanceolata* (Rasana): Chemical and biological potential of Rasayana herb used in traditional system of medicine. *Fitoterapia* 83(8): 1371–1385.

Tilley D. 2012. Plant Guide for Canadian horseweed (*Conyza anadensis*). USDA-Natural Resources Conservation Service, Aberdeen.

USDA 2012. Dr. Duke's Phytochemical and Ethnobotanical Databases: Plants with a chosen chemical. Eugenol *United States Department of Agriculture, Agricultural Research Service*, Washington, DC.

Vardhana R. 2007. Flora of Ghaziabad District. Shree Publishers and Distributors, New Delhi.

Viswanathan M.B., Kumar E.H.P.K. and Ramesh N. 2006. Ethnobotany of Kanis. Bishen Singh and Mahendra Pal Singh, Dehradun.

Wang L., Waltenberger B., Pferschy-Wenzig E.M., Blunder M., Liu X., Malainer C., Blazevic T., Schwaiger S., Rollinger J.M., Heiss E.H., Schuster D., Kopp B., Bauer R., Stuppner H., Dirsch V.M. and Atanasov A.G. 2014. Natural product agonists of peroxisome proliferator-activated receptor gamma (PPARdakua): a review. Biochemical Pharmacology S0006-2952(14): 424–429.

Whiting P.F., Wolff R.F., Deshpande S., Di N.M., Duffy S., Hernandez A.V., Keurentjes J.C., Lang S., Misso K., Ryder S., Schmidlkofer S., Westwood M. and Kleijnen J. 2015. Cannabinoids for Medical Use: A Systematic Review and Meta-analysis. *JAMA* 313(24): 2456–2473.

Yusuf M., Begum J., Hoque M.N. and Chowdhury J.U. 2009. Medicinal Plants of Bangladesh. Bangladesh Council of Scientific and Industrial Research Laboratories, Chittagong, Bangladesh.

Chapter 9

Fernery of Government Botanical Garden, Udhagamandalam

K. Anitha, R. Surendranath, K. Kayalvizhi and P. Rajasekar*

Department of Floriculture and Landscaping,
Tamil Nadu Agricultural University, Coimbatore,Tamil Nadu
**E-mail: anithasujay@gmail.com*

Abstract

Ferns are spore-producing plants under Phylum Filicinophyta (previously categorized under Pteridophyta) whose presence in the gardens gives wilderness with tenderness. Pteridophytes (ferns and fern allies) were once a principal component of terrestrial ecosystems dominating the Carboniferous landscape and are of great evolutionary significance. Ferns are popular horticultural plants and many species are grown in ornamental gardens or indoors and are estimated to have 6000 to 15,000 species with about 150 Genera and of great ecological significance. Ferns are used by people for ornamental purposes such as binding and decorative purposes as well as pot plants. Ferns in general provide an ecological service as bio-indicators for habitat health because of their sensitivity and specificity for temperature, humidity, soil type, moisture, pH, light levels, etc. Ferns thrive in a variety of habitats where regular flowering plants would often fail to dwell. The Government Botanical Garden, Udhagamandalm was established in 1848 and has a repository of a thousand species (both exotic and indigenous) of plants, shrubs, ferns, trees, herbal and bonsai plants and a fernery called W. C. McIvor fern house harboring ca. 127 species of ferns. But at present, the availability of fern species is around 30-35 and the fern anthology ranges from the delicate Adiantum sp, Nephrolepis sp, Platycerium bifurgatum, Anemia rotundifolia, Blechnum occidentale, Osmunda regalis, Micrasorum alternifolium, Asplenium sp to large sized tree fern (Alsophila sp). No doubt this fernery provides the garden an elegant look with the beautiful collection of ferns along with some fern allies like Selaginella sp and Ephedra sp. Reasons for the diminution of fern diversity are ubiquitous. Under such situations, in vitro and controlled environments of conservation strategies are the only panacea.

Keywords: *Garden, Udhagamandalm, Fernery, Fern, Spore*

Introduction

India is one of richest reservoirs of biological diversity in the world. It is the home for ferns, immediately capture the imagination of all who are fortunate enough to notice them and delight us with their aesthetic qualities and intrigue us with their unusual shapes and life histories. With their large, highly dissected and shiny green leaves, ferns are so visually appealing that many are sold as ornamentals. Most of the ferns of the carboniferous became extinct but some later evolved into our modern ferns. There are thousands of species in the world today. During the last five decades, ferns have faced a new kind of threat as humans have caused the destruction of natural habitats and unparalleled harvesting either for smuggling or medicinal uses (Pimm and Raven, 2000). Leaves of the South African fern *Rumohra adiantiformis* (Leatherleaf fern) are grown commercially for their ornamental value for flower arrangements. Horticultural research has resulted in hundreds of fern species being grown as ornamentals and successfully cultivated in greenhouses and fields throughout the world (Jones, 1987; Rickard, 2000; Hoshizaki and Moran, 2001; Mickel, 2003; Olsen, 2007).

Structure

Ferns and fern-allies are more complicated in structure. The leaf of the fern is frond which is divided into two main parts, the stipe (leaf stalk or petiole) and the blade (the leafy expanded portion of the frond). "Pinnate" blades are divided into leaflets (pinnae) and the ultimate division is called pinnule. Another type of division "pinnatifid" is one where the green leafy tissue isn't completely separated from the rachis but rather it spreads along the rachis.

Propagation

Miniature sacks or capsules like sporangia, that produce the dust like spores by which ferns are propagated. Several sporangia grouped together are called a Sorus. Most ferns would have their sporangia on the underside of the frond, arranged in an organized pattern usually associated with veins in the pinnule (leaf). Indusium acts as a protective covering for the sorus. The "seeds" of the ferns and fern allies are called spores.

Life Cycle of Ferns

Spores from the parent fall to the ground and find suitable moisture and light. The tiny single-celled organism starts to grow by cell division to form little green heart shaped plants or Prothallia (gametophytes).This is an independent plant with its own simple "root" system (rhizoids) to provide it with nutrients and water. The Prothallium then grows Antheridia or male organs and Archegonia or female organs on its underside.

Class	*Order*	*Family*	*No. of Genus*	*No. of Species*
Psilotopsida	Ophioglossales	Ophioglossaceae	3	87-92
	Psilotales	Psilotaceae	2	12
Equisetopsida	Equisetales	Equisetaceae	1	15

Class	*Order*	*Family*	*No. of Genus*	*No. of Species*
Marattiopsida	Marattiales	Marattiaceae	6	91-111
Polypodiopsida	Osmundales	Osmundaceae	4	19
	Hymenophyllales	Hymenophyllaceae	9	429+
	Gleicheniales	Gleicheniaceae	6	138
		Dipteridaceae	2	11
		Matoniaceae	2	4
	Schizaeales	Lygodiaceae	1	25
		Anemiaceae	1	120
		Schizaeaceae	2	35-40
	Salviniales	Marsileaceae	3	54-77
		Salviniaceae	2	16
	Cyatheales	Thyrsopteridaceae	1	1
		Loxomataceae	2	2
		Culcitaceae	1	2
		Plagiogyriaceae	1	15-20
		Cibotiaceae	1	10
		Cyatheaceae	5	558
		Dicksoniaceae	3	28
		Metaxyaceae	1	2
	Polypodiales	Lindsaeaceae	8	199-200
		Saccolomataceae	1	12
		Dennstaedtiaceae	11	189
		Pteridaceae	53	1104-1110+
		Aspleniaceae	2	735
		Thelypteridaceae	5	933
		Woodsiaceae	15	735
		Blechnaceae	9	244
		Onocleaceae	4	5
		Dryopteridaceae	33	1663-1685+
		Lomariopsidaceae	4	71
		Tectariaceae	10	294-297
		Oleandraceae	1	40
		Davalliaceae	5	665
		Polypodiaceae	62	1487- 1508+

Source: Smith *et al.* (2008). Suprageneric classification of ferns follows Smith *et al.* (2008). Generic classification and the estimates of genus and species counts in parentheses as given by A. R. Smith)

Diversity of Ferns in Ooty Botanic Gardens

The Ooty Botanical Gardens aka W. C. McIvor Fern house (Figures 9.1 and 9.2) enjoys a pleasant and temperate climate and receives an average rainfall of 140 cm mainly through the south-west monsoons and captures a massive area of 22 hectares with the spectacular Blue Nilgiris Mountain Range as the perfect backdrop setting. The Ooty Botanical Gardens houses thousands of species of indigenous and exotic trees, shrubs, ferns, bonsai and herbal plants. About 127 species of ferns were present in which, the availability of ferns is around 30-35 species.

Available Fern Species and their Description

Fern Species	*Common Name*	*Family*	*Description and Uses*
Adiantum paccotti (Figure 9.3) *Adiantum pedatum* (Figure 9.4) *Adiantum gracillimum* (Figure 9.5) *Adiantum williamsii* (Figure 9.6)	Maiden hair fern	Pteridaceae	Sori mostly marginal with false indusial Beautiful foliage is usually borne at the end of a relatively long, splendor polished black or purplish stalks Good as potted plants, indoor decoration and for hanging baskets Fresh leaves – bouquets, buttonholes Dried leaves – For greeting cards
Alsophila sp.	Tree fern	Cyatheaceae	Scaly tree ferns with abaxial sori Have a straight, tall stem or trunk The large leaves are borne at the apex of the trunk giving the plants a palm like appearance
Anemia rotundifolia (Figure 9.7)	Flowering fern	Anemiaceae	Sporangia on 2 to several, erect, basal pinnae The portion of the leaf which bears spores does not have any expanded leaf lamina and hence looks like an inflorescence Suitable for pot culture and hanging basket
Asplenium sp (Figure 9.8)	Bird's nest fern	Aspleniaceae	Clathrate rhizome scales, linear sori along veins Attractive species of ornamental ferns, either with entire leaves or with finely divided fronds
Blechnum occidentale (Figure 9.9)		Blechneceae	Sori along midvein, indusial open inwardly Have short stems and many have palm like leaves Leaves turn black and brown if watered from overhead
Davallia bullatta (Figure 9.10)	Squirrel's foot fern	Davalliaceae	Rhizomes long-creeping, petioles abscising at base Epiphytic and epipetric Some species resemble a hare's or a squirrel's feet

Fern Species	*Common Name*	*Family*	*Description and Uses*
			Suitable species for hanging baskets and for training on frames resembling birds and animals
Microsorum alternifolium (Figure 9.11)		Polypodiaceae	Sori round to oblong, exindusiate Leaves abscising often leaving phyllopodia Mostly epiphytic and epipetric Veins often areolate with included veins
Nephrolepis exceltata (Figure 9.12) *Nephrolepis tuberosa* (Figure 9.13) *Nephrolepis smithii* (Figure 9.14)	Sword fern	Lomariopsi-daceae	Creeping to climbing rhizomes, some epiphytes Suited for growing in hanging baskets and conservatories Produce slender runners
Osmunda regalis	Royal fern	Osmundaceae	About 128-512 spores per sporangium Stipules green, trilete spores Leaves dimorphic or with distinct fertile pinnae The fronds are feather like, plain or sometimes crested Grow as tall as 150 – 180 cm
Platycerium bifurcatum (Figure 9.15)	Staghorn fern	Polypodiaceae	Sori round to oblong, exindusiate Mostly epiphytic and epipetric Veins often areolate with included veins Fronds are divided at the margin like the horns of stages Grown like orchid in logs, preferably supported by fibres such as coconut and sphagnum moss Suitable for hanging baskets
Polystichum angular		Dryopterida-ceae	Petioles with 3 or more vascular bundles Sori mostly round to oblong Moisture loving hardy ferns which grow better in semi shade or shade, though tolerate enough sunlight Useful in flower arrangements

Conclusion

The scope of ecological studies of ferns is also changing, widening to include large-scale spatial and temporal dynamics as well as social and economic influences. While research on the individual species provides the fundamentals, it is within the larger context of global ecology that the role of ferns may prove vital to our future, given their long and successful history of adaptation to environmental change shapes and life histories. When examined more closely, ferns also bear strong resemblances

Figure 9.1 Figure 9.2 Figure 9.3

Figure 9.4 Figure 9.5 Figure 9.6

Figure 9.7 Figure 9.8 Figure 9.9

Figure 9.10 Figure 9.11 Figure 9.12

Figure 9.13 Figure 9.14 Figure 9.15

to seed plants. This awareness and engagement with ferns by humans is beneficial for the survival of ferns. However, the high diversity of ferns on tropical mountains, on remote islands and in forest under stories worldwide makes ferns susceptible to the increasing human impacts on those ecosystems. Fern hot spots should be identified and conservation of rare ferns should become a priority for conservation advocacy groups. The prohibition of international trade in living tree ferns by CITES in 2000 was an important step toward appropriate recognition of the vulnerability of ferns when wild populations are harvested without sustainable management. Cultivation and propagation can help the recovery of some populations and assist restoration efforts but these approaches are not sufficient to replace conservation of natural areas. Because ferns are dynamic participants in many habitats, represent ancient and modern lineages of plants, and simultaneously resemble and differ from other plants, we hope and expect that the study of fern ecology will be a high priority for future generations. The successful conservation of ferns and lycophytes requires a number of actions such as improving knowledge of their abundance and distribution patterns and conducting field studies to investigate habitat specificity, population dynamics, genetic diversity and any intrinsic factors or environmental disturbances that affect their survival. The propagation of endangered ferns in tissue culture is an additional tool for conservation that also can satisfy the demand for ornamental ferns without damaging natural populations.

References

Hoshizaki, B. J. and Moran, R. C. 2001. The Fern Grower's Manual, revised and expanded edition. Portland, OR, USA: Timber Press.

Jones, D. L. 1987. Encyclopedia of Ferns. Portland, OR, USA: Timber Press.

Mickel, J. T. 2003. Ferns for American Gardens. Portland, OR, USA: Timber Press.

Olsen, S. 2007. Encyclopedia of Garden Ferns. Portland, OR, USA: Timber Press.

Pimm, S. L. and Raven, P. 2000. Extinction by numbers. *Nature*, 403, 843–5.

Rickard, M. 2000. The Plantfinder's Guide to Garden Ferns. Portland, OR, USA: Timber Press.

Smith, A. R., Pryer, K. M., Schuettpelz, E., *et al.*, 2008. Fern classification. In Biology and Evolution of Ferns and Lycophytes, ed. T. A. Ranker and C. H. Haufler. Cambridge, UK: Cambridge University Press, pp. 417–67.

Chapter 10

Conservation of Orchids at TNAU Botanic Gardens

M. Kannan and P. Ranchana

Department of Floriculture and Landscaping, Horticultural College and Research Institute, Tamil Nadu Agricultural University, Coimbatore – 641 003, T.N.
E-mail: kannanflori@gmail.com

Abstract

Orchids are the one of the most important genera in the plant family which possesses wide range of habitat (from terrestrial to epiphytic) with incredible diversity. They constitute an order of royalty in the world of ornamental plants and they have immense horticultural importance. They play a vital role to balance tne natural ecosystem and are valued mostly for their exquisite flowers, which are available in a vast array of colours (tints of blue, yellow, white, orange and red to almost black), form, size, shape, amazing long-lasting flowers which occupies top postion among all the flowering plants and is valued both for cut flower production and as potted plants. Curious and attractive blooms of the species and hybrids of tropical orchids grown in this orchid house of TNAU notanic garden are a feast to the eyes of visiting tourists. Moreover, orchids are the major players in the multibillion dollar floriculture trade of the world. Among the various components present in the TNAU botanic garden, orchid house is a protected structure for growing and displaying orchids which require sheltered growing conditions. They include Dendrobium, Epidendrum, Phalaenopsis, Spathoglottis, Vanda, etc. Among them Dendrobium, Epidendrum, Phalaenopsis and Vanda are epiphytic in nature and Spathoglottis is of terrestrial nature. Now-a-days, these orchids are getting popular as houseplants because of their affordability, easy care and maintenance to deliver a colourful treat for the eyes. Tropical orchids are now adaptable for growing in pots.

Keywords: *Orchids, Colours, Habitat, House plant, Cut flower, Pot plant.*

Introduction

Orchids are distinctive plants and highly priced in the international floricultural trade due to their intricately designed spectacular flowers with brilliant colours, delightful appearance, myriad sizes, shapes, forms and long lasting qualities. Orchid cut flowers have emerged as the leader in the international market and have immensely contributed to the economy of several developed and developing countries. Orchids are excellent for garden and can be grown in beds, pots, baskets, split hollows of bamboo pieces, *etc*. They remain the only cut flower crop grown as a pot plant in most parts of the world due to the epiphytic nature of the commercially cultivated orchids.

Thailand, Singapore and Malaysia are major exporters of orchid flowers and plants. The Netherland is the single largest producer and exporter of temperate orchids, while Thailand and Hawaii are world leaders in production of tropical orchids. Thailand has a long history of orchid trade, especially for export. It is estimated that 54 per cent of the orchids produced are exported and the rest 46 per cent consmed in the local market. The biggest importers for Thailand orchids are Japan (30 per cent), USA (22 per cent), Italy (10 per cent) and China (9 per cent). During 2012-13, Dendrobium contributed 94.7 per cent of total orchid cut flowers followed by Mokara (3.1 per cent), Aranda (0.9 per cent), Oncidium (0.7 per cent), Aranthera (0.5 per cent) and Vanda (0.1 per cent), key procucts are whole stems (88 per cent), loose blooms (8.4 per cent) and garlands (3.5 per cent).

Classification of Orchids Based on Type of Growth

Orchids exhibit following two types of growth:

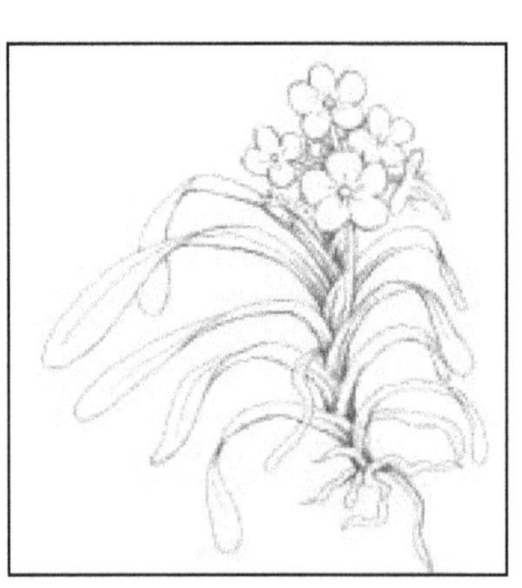

Monopodial

Sympodial

Monopodials

Monopodials are plants with single non-branching stem, which grows up year after year. Leaves, roots and inflorescences are produced from the nodes, along the entire length of the stem. The roots absorb moisture and nutrients from the air. In general, the monopodial (meaning 'single footed') has a vertical growth. **Examples:** *Vanda, Phalaenopsis* (moth orchid) and *Arachnis* (scorpion orchid/spider orchid).

Vanda

This belongs to monopodial group. The members of this group are generally sun loving and robust with a wide range of colourful, beautiful shaped flowers which are heavy substances and long lasting. Based on the shape of the leaves, there are two types of vandas, the strap leaved and terete (pencil-like) leaved. Intermediate types, like semi-terete and quarter terete types also exist. Leaves grow alternately on either side of the main stem. Aerial roots and inflorescences develop from the leaf axil. It produced flowers on a lateral inflorescence. Most of the flowers exhibit yellow-brown colour with brown markings, but also appear in white, green, orange, red and burgundy shades. **Examples:** Miss Joaquim, John Clubb, Wirat Uchida, Bill Sutton, Ellen Nos, Emily Notley, Evening Glow, Honomu, Honolulu, Hilo Blue, Alice Blue, Josephine van Brero Crosses, Eman van Deventer Crosses, Walter Gumae and Onomea.

Sympodial

The sympodial plants have a rhizome which grows horizontally and produces new growth. Thus, a well-developed sympodial plant contains a clump of shoots of varying age and size. In some genera of this group (*e.g. Cattleya, Dendrobium*) the stems are thickened and are called 'pseudobulbs'. The new growth originating from the base of a sympodial orchid is called a 'lead', which indicates the direction of growth of the plant. The sympodials (meaning 'united feet') have a horizontal growth. **Examples:** *Cattleya, Dendrobium, Oncidium* (dancing lady), *Cymbidium, Paphiopedilum* (lady's slipper orchid).

Cattleya

There are two types based on leaf types, the unifoliate which produces single leaf from each pseudobulb and bifoliate which produces two or even three leaves. The pseudobulbs and the flowers are usually larger in size. They require partial shading under tropical conditions. There are more than 50 species. **Examples:** Bob Betts, Bow Bells, Bowrevel, Chickamauga, Diane Salo, Doty Sykora, Empress Bells, Margaret Stewart and White Christmas are noticeable varieties.

Dendrobium

This is the second largest genus of orchids consisting about 1340 species. They produce pseudo bulbs or slender canes out of the underground rhizomes. The members of this genus are either epiphytes or lithophytes. They have an erect, cane-type pseudobulb and have flowers with relatively long vase life. They prefer partial shade and high humidity and several commercial varieties are found to perform extremely well under tropical conditions. They need very good air circulation.

Examples

- **Dark Pink:** Sonia 17, Sonia 28, Deang Suree, JO 17, Nia Rose, Uniwai Prince
- **Light Pink:** Intuwong, Sakura Pink, Miss Teen, Anna, Bankok Blue, Pahoo, Yoko
- **White:** Pure White 4N, Jack Dorean (Sanan White), Aliga, Blushing, Big White Sunshine, Emma White, Kasem White
- **Yellow:** Kasem Gold, Jiad Gold, Fatima, Jade, Mary Mark
- **Violet:** Violet, Blueberry
- **Bicoloured:** Dark Sonia (D- BOM), Sonia (Bombay or BOM), Fatima Red,Misteen, Putiwa
- **Green:** Mary Trouse
- **New Pink:** It bears purple florets with deep purple petals and light purple sepals. Petals are larger than sepals. Lip is deep purple in colour with feathery outgrowth and white towards centre. Spikes are 30 to 33 cm long, bearing around six florets.
- **Emma White:** Creamy white coloured flowers. Creamy white sepals, petals and lip. Petals are larger than sepals. Length of spike ranges from 1½ to 2 feet with seven florets.
- **Kasem White:** Florets are creamy green in colour. Sepals are white coloured and have petals with green tinge. Lip is creamy green with feathery longitudinal striations. The spike is around 38-40 cm long with six florets.
- **Sonia 17:** White and purple coloured flowers. Sepals creamy white with purple marking and petals purple in colour. Lip light purple with cream coloured.
- **Sonia 28:** Purple and white coloured flowers. Deep purple petals and lip with white colouration towards the centre. Sepals white with a purple tinge. Petals a little larger than sepals. Length of spike ranges from 35 cm to 43 cm with twelve florets.
- **Jay Swee King x Jacquline Concert:** It produces deep purple coloured flowers. Length of the spike is 50 cm with twelve florets.
- **Jacquline Thomas:** Light greenish white flowers with purple lines in the lip. Length of the spike is 36 cm with five florets.
- **Madam Vipor:** Flowers are light green with purple lip. Petals and sepals also light green in colour. Length of the spike is 20 cm with four florets.
- **Sakura Pink:** Light purple coloured flowers. Light purple coloured petals, sepals and lip. Petals and sepals are of almost equal size. Length of the spike is 24 cm with seven florets.

Oncidium

Oncidium is often called 'Dancing Lady' because of the graceful doll like flowers. About 750 species are listed in this genus. Some oncidiums have prominent pseudobulbs, usually flattened, while others have none at all. The temperature and light requirements are similar to those of *Cattleya* and *Dendrobium*. Gower Ramsey, Golden Shower, Josephine, St. Anne, Tiny Tim and Wikike Sunset are few important varieties.

Classification Based on Growth Habitat

Orchids can be divided into two broad groups based on habitat.

Terrestrial Orchids

Grown on the ground (*e.g. Spathoglottis plicata, Cymbidium, Paphiopedilum*).

Epiphytic Orchids

Have aerial roots and grow on trees (*e.g. Vanda, Dendrobium*)

Temperature Requirements of Orchids

Category	*Temperature (° C)*	
	Day Temperature	*Night Temperature*
Cool orchids	15 – 21	10 - 12.5
Intermediate	18 – 21	15.5 – 18
Warm orchids	21 – 29	18 – 21

Propagation

Stem Cuttings

The main method of propagation is through stem cuttings. About 40-50 cm long top cuttings with at least two well-developed aerial roots are ideal. Intermediate cuttings can also be used. If smaller cuttings (2-3 nodes) are used, the time taken for flowering will be longer.

Division

This method involves division of large clumps into smaller units. This is the most common method of propagation in sympodial orchids. Care should be taken to see that each unit has at least 4-5 shoots, including the old ones. Bigger the size of division, faster the establishment and earlier the flowering. This method is especially suitable for *Cattleya, Dendrobium, Cymbidium, Epidendrum etc*. Care should also be taken not to damage the roots.

Off-Shoots or Keikis

Orchids like *Dendrobium* sometimes produce small plants with roots at the nodes of pseudo bulbs. These are called *'keikis'* meaning 'babies'. These, when sufficiently grown, are to be separated carefully from the mother plant and potted independently.

Tissue Culture

Various plant parts like shoot tips or meristems, leaf segments, stem segments, floral parts, aerial roots, *etc.* have been used for tissue culture of orchids. Meristem culture is more popular and is extensively used for the commercial propagation of orchids. The widely used nutrient media are Knudson's C medium, Vacin and Went's medium and Murashige and Skoog's medium, *etc.* Additives like coconut water (15 per cent) banana pulp (10 per cent) *etc.* are found beneficial for the promotion of shoots.

Special Horticultural Practices

Containers which have many holes on their sides are suitable since they help in maintaining a porous medium and well aerated conditions. Some of the most popular containers presently used by the tropical orchid gowers are pots, baskets, wooden logs, tree fern, coconut husk, *etc.* They are either placed on the benches or hanged from top. Claypots are generally preferred to the plastic pots. The pot size may vary from few inches to 20 inch pot depending on the size of planting material. Climbing orchids and terrestrial orchids can be generally planted on the ground. Epiphytic orchids are grown in some type of containers or supports.

Growing Media for Epiphytic and Terrestrial Orchids

Orchids are the major cut flowers grown mostly as a pot plant. Potting medium differs with the type of orchids. Monopodial epiphytic orchids perform best when grown in chunks of hardwood charcoal, while sympodial epiphytes grow better in osmunda tree fern fibre. Apart from the above media, several other media are used in different combinations for monopodial and sympodial epiphytes like barks of trees like pine, fir, coconut husk, overburnt bricks, solid polyurethane foam, coarse vermiculite (0.63 cm to 2.5 cm), polystyrene, Styrofoam chips, calcinated clay, pumic chips, polypodium fibre and sphagnum moss. Clay pots with charcoal are ideal for large scale production. Now-a-days blocks prepared from coconut husk (cocopeat blocks) are being used for commercial production.

Terrestrial orchids perform well in organic rich porous compost mixture. Combinations of media are made with loamy soil, leaf mould, river sand, dust free bark preparation, charcoal dust, tree fern fibre, eucalyptus leaf litter, dolomite soil, saw dust, volcanic soil, old cowdung, finely broken crocks, perlite, milled peanut shell, peat moss and sphagnum moss depending on the species and genera. Aranda and *Arachnis* are successfully grown in organic rich soils in ground beds.

Benches for Cultivation

For making the stages to place the orchids, concrete piles with a thickness of 1.5 x 1.5 inch are used. The piles of the stages for placing the orchids rise above the ground about 60 to 70 cm, The stage is 1.20 m wide and the width between two piles is 80 cm, 1.20 metre-long concrete beams are placed between two piles. When not using concrete beams, steel pipes will be best alternative. The laying of waterpipes and branch them into different directions makes connecting the hose and watering the plants more convenient.

Suitable Environment for Growth and Flowering

Temperature requirements of orchids vary to a great extent, since they are found in a wide range of climatic situations in their natural habitats. For cultivation of Dendrobiums, a minimum temperature of 15°C to a maximum of 33°C is required. The EC of water should ber 0.5 ms/cm. Most orchids like good light conditions, for example, *Aerides, Cattleya, Epidendrum, Rhynchostylis* and *Vanda.* Some of the orchids are light-loving while others are shade loving. Orchids such as *Arachnis, Vanda etc.* prefer open sunlight conditions to grow and bloom. While growing these types of orchids care should be taken to ensure that the plants receive adequate amount of light round the year.

Certain orchids prefer partially shaded conditions for their growth and development *e.g. Dendrobium, Phalaenopsis, Cattleya, Oncidium.* The extent of shade requirement may vary among the genera. Light can be regulated using the right type of shade net. In general, a grade of 50 per cent shade would be adequate for most of the shade loving orchids.

Different species of orchids exhibit their sensitivity to day-length in various ways and yet small changes in day-length are apparently critical for growth and flowering of these plants. Maintenance of proper humidity in the orchid house is most important. Evergreen species of *Dendrobium* prefer more humidity (50 to 70 per cent) than deciduous species. The moderate humidity range is 40-70 per cent for *Paphiopedilum* and *Cyperpedium,* 70-75 per cent for *Aerides, Phalaenopsis, Rhyncostylis* and *Vanda* and 40-55 per cent for *Cattleya* and *Laelia.*

Nutrition

Application of 0.2 per cent of NPK @ 10:20:10 is recommended at weekly intervals. Spray application of cow dung and oilcake solution and micronutrients gave better results in several epiphytic orchids. Use of coated and slow release fertilizers, for example, Osmocote is most suitable for terrestrial orchids.

Harvesting

Most orchids are harvested when two or three buds are still unopened. The weather conditions prevailing at the time of harvesting influence the exact stage of picking. During the warm weather, the blooms can be harvested at an earlier stage of picking. Flowers are generally harvested twice a week during peak production period and once a week during low periods. Harvesting should preferably be done in the evening. The knife used for harvesting may be dipped in an antibiotic solution to prevent disease transmission.

Yield

On average, tropical orchids produces 2-3 marketable spikes in the second year and produces 5-6 spikes from third year onwards.

Grading

It is done mainly on length of the flower spike, floret number and size and arrangement of florets on the spike.

Post-harvest Handling of Orchid Cut Flowers

Cattleya

Stage of harvest: 3-4 days after full opening of florets

Storage: Flowers are stored wet at 8 to 10°C for 10 to 14 days

Vase life: 14 days

Dendrobium

Stage of harvest: When all florets open

Storage: Flower spikes can be wet stored at 5 to 7°C for 10 to 14 days

Vase life: 14 - 21days, floral preservatives are not very effective, but some improvement can be seen if kept in a holding solution of 2 per cent sucrose and 200 ppm 8-HQC. Other preservative solutions are 4 per cent sucrose plus 30 ppm $AgNO_3$, 4 per cent sucrose plus 30 ppm $AgNO_3$ plus 225 ppm 8-HQC. Foliar application of Aluminium chloride (500 ppm) or Ammonium molybdate (100 ppm) or Boric acid (1000 ppm) enhances the vase life.

Oncidium

Stage of harvest: When most of the florets on the spray are fully opened

Storage: 7 to 10°C for 10 to 14 days

Vase life: 10 – 14 days. Boric acid 250 ppm in vase solution increases the vase life of cut spike. Four per cent sucrose plus 100 ppm acetylsalicylic acid or 100 ppm 8-HQC increases the percentage of floret opening.

References

AICRP Annual Report 2010 – 11. Regional research station (Hill Zone), Uttar Banga Krishi Viswa Vidyalaya, Kalimpong.p.400.

Arumugam, T. and Jawaharlal, M., 2004. Effect of shade levels and growing media on growth and yield of *Dendrobium* orchid cultivar Sonia-17. *Journal of Ornamental Horticulture,* 7(1): 107-110.

Bose, T.K., 1978. Commercial Flowers. Naya Prakash, Calcutta,: 638.

De, Khan, A., Kumar, R. and Medhi, R.P., 2014. Orchid Farming- A Remunerative

Approach for Farmers Livelihood. *International journal of scientific research,* 3(9).

Hew, C. S. 1987. The effect of 8-Hydroxyquinoline sulphate, Acetylsalicylic acid and Sucrose on bud opening of *Oncidium* Flowers. *J. Hort. Sci.,* 62(1):75-78.

Ketsa, S. and D. Amutiratana. 1986. Effect of Sucrose, Silver nitrate and 8-Hydroxyquinoline sulphate on postharvest behaviour of *Dendrobium 'Pompadour'* flowers. Proc. Sixth ASEAN Orchid Congress. Bangkok, Thailand, p.124-129.

Kumar, 1992. Potting media and post transplantation growth of *Dendrobium* hybrid seedlings. *Journal of the Orchid Society of India,* 6(1-2): 131-133.

Naik, Raja, M. and K, Ajithkumar., 2014. Influence of Plant Growth Promoters and Growing Systems on Flowering of *Dendrobium* cv. Earsakul. *Agric. Technol,* 1: 25-35.

Nair, S.A. and Sujatha, K., 2010. Effect of varying levels of foliar nutrients on round the year production and quality of *Dendrobium* cv. Sonia 17. *Journal of Ornamental Horticulture,* 13(2): 87-94.

Sabina, G.T., 1996. Performance of selected orchids under varying light regimes, culture methods and nutrition. Ph.D thesis submitted to Kerala Agril. Univ., Thrissur.

Shankar, K.S., Manivannan, K. and Kamalakannan, S., 2003. Effect of potting media and nutrients on flowering of *Dendrobium* hybrid 'White Fairy'. Nutritional symposium on Recent Advances in Indian Floriculture: 97 - 100.

Sheehan, T., 1961. Effects of nutrition and potting media on growth and flowering of certain epiphytic orchids. Amer. *Orchid Soc. Bull,* 30(4): 289-292.

Swapna, S. 2000. Regulation of growth and flowering in *Dendrobium* cv. Sonia 17. Ph.D. thesis, Kerala Agricultural University, Thrissur, Kerala, p.235.

Usha, D.N., 2000. Influence of nitrogen and harmones on growth and development of *Dendrobium* cv Sonia - 17. M.Sc thesis: 105-107.

Yadav and Bose., 1989. Commercial Flowers. Naya Prakash, Calcutta: 208.

Chapter 11

Conservation of Trees and Shrubs in GNDU Botanical Garden

Gurveen Kaur, Jaskirat Kaur and Avinash Kaur Nagpal

Department of Botanical and Environmental Sciences,
Guru Nanak Dev University, Amritsar – 143 005
E-mail: bajwagurveen@gmail.com, jaskiratk888@yahoo.com,
avnagpal@rediffmail.com

Abstract

Plant diversity is a sum total of various forms of plants on earth. Over Exploitation or excessive use of species, pollution, natural disasters etc. are the factors resulting in increasing rate of extinction of different plant species. There is a need of conservation of plant diversity. This necessitates that we focus our attention to conserve different plant species available as of today. Botanical Gardens play an important role in conservation of plant diversity. Botanical Garden is a place where different types of plant species are conserved for the purpose of scientific research, education and display etc. Guru Nanak Dev university Botanical Garden was established in the year 1975, covers an area of 25 acres and has a collection of nearly 500 plant species including annuals, cacti, medicinal and seasonal plants. The plants are arranged according to Bentham and Hooker's system of classification.

Keywords: *Trees, Shrubs, Botanical Garden, Conservation.*

Introduction

Biodiversity referring to millions of different life forms present on this earth forming the core of our life support system, is so important for sustainable existence of all life forms on this earth that removal of even a single species can upset the balance and disrupt or destroy the ecosystem threatening all forms of life within it. During the last few decades we are witnessing a rapid increase in extinction of plants species even those which we have not yet been able to identify due to ever increasing human population, urbanization, deforestation, habitat destruction, over exploitation *etc*. Biodiversity is being significantly reduced by a wide range of human activities; however, habitat destruction is the main threat (Dompka 1996; Mittermerier *et al.*, 1999; Thompson and Jones 1999).

Sincere efforts at global and local level are required to minimize as well as remediate such negative impacts on biodiversity. Concerns over the accelerating loss of biodiversity have also aroused the attention of governments worldwide who have initiated efforts for formulating biodiversity friendly policies to improve economy and alleviate poverty by conservation of their living resources. In these on-going multi-prolonged efforts to halt species extinction and to promote the conservation of our plant genetic resources, Botanical Gardens play a pivotal role. Botanical Gardens are the institutions holding documented collections of living plants for the purpose of scientific research, conservation, display and education (http://www.anbg.gov.au/anbg/what-is-a-botanic-gardens.html). According to Almond (1993) "A garden is an ever changing museum of living plants". A botanic garden is a garden which maintains collection of living plants which are well labeled and arranged on some scientific basis (Chakravorty and Mukhopadhyey, 1990). The traditional role of botanic gardens has been to identify, classify and grow plant species collected from different parts of the world for conservation, display and education (Balick, 1986). It is expected that the existing Botanic Gardens have to balance their traditional role of being botanical warehouses displaying a range of species for taxonomic research and the present role of growing and conserving variety of threatened species (Mander, 1994).

The GNDU Botanical Garden, Amritsar is an integral part of the Department of Botanical and Environmental Sciences. It was established in the year 1975 and is spread over an area of 25 acres. The nature of the soil is porous and saline. It has a collection of nearly 500 plant species including annuals, cacti, ferns, rare, threatened, and endangered as well as medicinal plant species. Trees, shrubs, herbs, climbers *etc*. are arranged section wise as per Bentham and Hooker's system of classification. The important features of the GNDU Botanical Garden include green house, poly house, screen house, fern house, herbal garden, lily pool, vermicomposting unit *etc*.

Identification of Trees and Shrubs

The Botanical garden of Guru Nanak Dev University has been divided into different sections as per the Bentham and Hooker's system of classification. Figure 11.1 shows the satellite image of the GNDU Botanical Garden from Google earth (http://earth.google.com). Figure 11.2 shows the map layout of the GNDU Botanical Garden prepared by tracing various paths in the Google image which divides the Botanical Garden into different sections including Coniferae, Cycadae, Monocotyledonae, Monochlamydae, Polypetalae and Gamopetalae. Apart from these there are also sections including Arboretum, Bambusetum, Teak plantation, Conservatory, Herbal garden, Fern house *etc*.

Figure 11.1: Satellite Image of GNDU Botanical Garden.

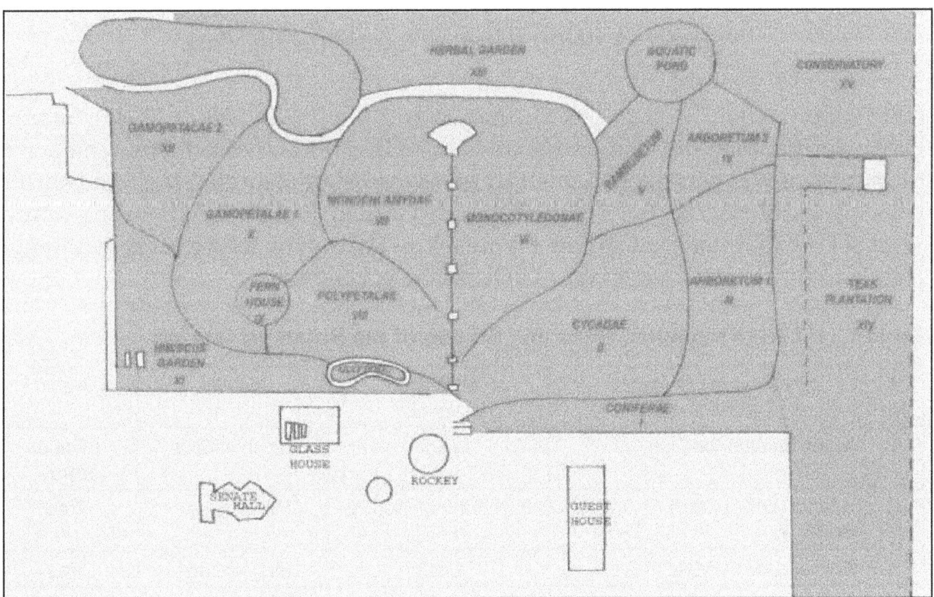

Figure 11.2: Layout of GNDU Botanical Garden.

Various trees and shrubs growing in different sections were identified by studying their botanical characters (vegetative and reproductive) and by comparing with the herbarium specimens of the Department and consultation of the works of Hooker (1885); Roxburg (1971); Stewart and Brandis (1972); Bamber (1976); Talbot

Figure 11.3: GNDU Botanical Garden Central Vista.

(1976); Stewart (1977); Nair (1978); Sharma and Bir (1978); Sabins (1986); Sharma (1990). Information regarding vegetative as well as floral/reproductive characters of each plant was compiled. Table 11.1 gives the list of identified trees and shrubs growing in different sections of the Botanical Garden. Figure 11.3 shows the central vista of GNDU Botanical Garden. Figures 11.4–11.10 show the pictures of different trees and shrubs growing in the GNDU Botanical Garden.

Table 11.1: List of Identified Trees and Shrubs of the Botanical Garden

Sl. No.	*Name of the Plant/ Botanical name*	*Common Name*	*Family*	*Habit*
1	*Abrus precatorious* Linn.	Rati	Papilionaceae	Climber Shrub
2	*Acacia auriculiformis* A.Cunn. ex Benth..	Auri, Earleaf, Acacia	Mimosaceae	Tree
3	*Acacia catechu* Willd.	Khair	Mimosaceae	Tree
4	*Acacia nilotica* Delile	Kikar	Mimosaceae	Tree
5	*Acalypha wilkesiana* Mull.Arg	Acalypha	Euphorbiaceace	Shrub
6	*Acer oblongum* Wall.ex.DC	Pangoi	Aceraceae	Tree
7	*Adathoda vasica* Nees	Arusha	Acanthaceae	Herb

Sl. No.	*Name of the Plant/ Botanical name*	*Common Name*	*Family*	*Habit*
8	*Aegle marmelos* Correa ex Roxb.	Bel	Rutaceae	Tree
9	*Agathis robusta* F. M. Bailey	Queensland kauri	Araucariaceae	Tree
10	*Albizzia lebbeck* Benth	Siris	Mimosaceae	Tree
11	*Albizzia procera* Benth.	Safed Siris	Mimosaceae	Tree
12	Aloe Vera Mill.	Kavar gandal	Liliaceae	Herb
13	*Alstonia scholaris* R.Br.	Chaitan tree, Saptaparni	Apocynaceae	Tree
14	*Anthocephalus cadamba* Miq.	Kadamba	Rubiaceae	Tree
15	*Anthocephalus* sp.	Anthocephalus	Rubiaceae	Tree
16	*Araucaria cunninghami* D. Don	Hoop pine, Monkey puzzle	Araucariaceae	Tree
17	*Bambusa arundinaceae* Willd	Bans	Poaceae	Tree
18	*Bambusa ventricosa* McClure	Buddha Bamboo	Poaceae	Tree
19	*Bambusa vulgaris* Schrad.	Common Bamboo	Poaceae	Tree
20	*Banisteria laurifolia* L.	Banisteria	Malphiginaceae	Shrub
21	*Bassia latifolia* Roxb.	Mahua	Amaranthaceae	Tree
22	*Bauhinia galphinii* N.E.Br	Pride of de Kaap	Caesalpiniaceae	Shrub
23	*Bauhinia purpurea* Linn.	Khairwal	Caesalpiniaceae	Tree
24	*Bauhina sp.*	Kachnar	Caesalpiniaceae	Tree
25	*Bauhinia tomentosa* Linn.	Kachnar	Caesalpiniaceae	Tree
26	*Bauhina vahlii* Wight and Arn.	Maljan	Caesalpiniaceae	Climber
27	*Bauhinia variegata* Linn.	Kachnar	Caesalpiniaceae	Tree
28	*Bignonia undulata* Sm.	Rugtrora	Bignoniaceae	Climber
29	*Biota orientalis* Endl.	Thuja	Cupressaceae	Shrub
30	*Bombax ceiba* L.	Semul	Bombacaceae	Tree
31	*Bougainvillea glabra* Choisy	Bougainvillea	Nyctaginiaceae	Shrub
32	*Butea monosperma* (Lam.) Kuntze	Dhak	Papilionaceae	Tree
33	*Callistemon lanceolatus* DC.	Bottle brush	Myrtaceae	Tree
34	*Canna indica* L.	Canna	Cannaceae	Herb
35	*Capparis aphylla* Roth	Kair	Capparidaceae	Shrub
36	*Caryota urens* L.	Mari, Kittul	Arecaceae	Palm
37	*Cassia biflora* L.	Biflora	Caesalpiniaceae	Tree
38	*Cassia fistula* L.	Amaltas	Caesalpiniaceae	Tree
39	*Cassia glauca* Lam.	Twin flowered cassia	Caesalpinaceae	Tree
40	*Cassia javanica* L.	Javanica	Caesalpiniaceae	Tree
41	*Cassia siamea* Lamk.	Kassod tree	Caesalpiniaceae	Tree
42	*Casuarina equisitifolia* L.	Jangli saru	Casuarinaceae	Tree
43	*Cedrela toona* Roxb.	Tun, Mahanim	Meliaceae	Tree

Sl. No.	*Name of the Plant/ Botanical name*	*Common Name*	*Family*	*Habit*
44	*Cestrum nocturnum* L.	Raat ki Rani	Solanaceae	Shrub
45	*Chukrasia tabularis* A.Juss.	Chickrassy	Meliaceae	Tree
46	*Cinnamomum camphora* (L.) Burm.f.	Kapur	Lauraceae	Tree
47	*Citrus jambhiri* Lush.	Jatti khati	Rutaceae	Tree
48	*Citrus limon* (L.) Burm.f.	Baranimbu, Jambira	Rutaceae	Tree
49	*Clematis paniculata* L.	Sweet autumn Clematis	Ranunculaceae	Climber Shrub
50	*Clerodendrone inerme* (L.) Gaertn.	Lanjai, Sangkupi	Verbenaceae	Shrub
51	*Combretum paniculatum* Vent.	Forest flame Creeper	Combretaceae	Climber
52	*Cordia dichotoma* Forst.f.	Lasora	Boraginaceae	Tree
53	*Crataevea religiosa* Hook.f. and Thoms	Barna, Bilasi	Capparidaceae	Tree
54	*Crinum album* Herbert	Natal lily	Amaryllidaceae	Herb
55	*Cryptostegia grandiflora* R.Br.	Rubber vine	Asclepiadaceae	Climber
56	*Cupressus semipervirens* L.	Sara, Mediterranean cypress	Pinaceae	Shrub
57	*Cycas circinalis* L.	Jangli-madan-mast-ka-phul	Cycadaceae	Shrub
58	*Cycas revoluta* Thunb.	King sago palm	Cycadaceae	Shrub
59	*Dalbergia sissoo* Roxb.	Shisham	Paplionaceae	Tree
60	*Delonix regia* Rafin.	Gulmohar	Caesalpiniaceae	Tree
61	*Dendrocalamus strictus* (Roxb.) Nees.	Calcutta Bamboo	Poaceae	Tree
62	*Dillenia indica* L.	Chalta	Dilleniaceae	Tree
63	*Diospyrous scadens* L.	Diospyrous	Ebenaceae	Tree
64	*Dombeya acerifolia* Baker	Maple leaved Dombeya	Malvaceae	Shrub
65	*Dracaena australis* G. Frost	Cabbage palm	Liliaceae	Shrub
66	*Emblica officinalis* Gaertn.	Amla	Phyllantaceae	Tree
67	*Erythrina indica* Willd.	Dapad	Papilionaceae	Tree
68	*Eucalyptus longifolia* Link and Otto	Safeda	Myrtaceae	Tree
69	*Ficus benghalenesis* L.	Banyan	Moraceae	Tree
70	*Ficus benjamina* L.	Weeping Fig	Moraceae	Tree
71	*Ficus elastica* Roxb.	Rubber tree	Moraceae	Tree
72	*Ficus glomerata* Roxb.	Gular, Umar	Moraceae	Tree
73	*Ficus infectoria* Roxb.	Kahimal	Moraceae	Tree
74	*Ficus religiosa* L.	Pipal	Moraceae	Tree
75	*Ficus trigona* L.f.	Common fig	Moraceae	Tree

Sl. No.	Name of the Plant/ Botanical name	Common Name	Family	Habit
76	*Gardenia jasminoides* J. Ellis	Gandharaj	Rubiaceae	Shrub
77	*Gmelina arborea* Roxb.	Gambhar	Verbenaceae	Tree
78	*Gmelina asiatica* L.	Badhara, Nag-phul	Verbenaceae	Shrub
79	*Grevillea robusta* R.Br.	Silver Oak	Proteaceae	Tree
80	*Grewia asiatica* L	Phalsa	Tiliaceae	Tree
81	*Hamelia patens* Jacq.	Scarlet Bush	Rubiaceae	Shrub
82	*Hemerocallis sp.*	Hemerocallis	Liliaceae	Herb
83	*Hiptage madablota* Gaertn	Madhmalti	Malpiginaceae	Climber
84	*Holmskiodia sanguinea* Retz.	Chinese hat	Verbenaceae	Climber
85	*Hypericum chinensis* L.	Hypericum	Hypericaceae	Shrub
86	*Ixora chinensis* Lam.	Ixora	Rubiaceae	Tree
87	*Jacranda mimosifolia* D. Don	Blue trumpet tree	Bignoniaceae	Tree
88	*Jasminum humile* L.	Yellow jasmine	Oleaceae	Shrub
89	*Jatropha curcas* L.	Jangli- arandi	Euphorbiaceae	Shrub
90	*Jatropha integerrima* Jacq.	Fire-Cracker	Euphorbiaceae	Shrub
91	*Juniperus chinensis* L.	Chinese Juniper	Pinaceae	Shrub
92	*Kigelia pinnata* DC.	Common Sausage tree	Bignoniaceae	Tree
93	*Koelreuteria paniculata* Laxm.	Golden rain Tree	Sapindaceae	Tree
94	*Lagerstroemia indica* L.	Pharash, Common Crape myrtle	Lythraceae	Shrub
95	*Lawsonia inermis* L.	Mehndi, Heena	Lythraceae	Shrub
96	*Ligustrum indicum* (Lour.) Merill	Indian Privet	Oleaceae	Tree
97	*Livistonia chinensis* R.Br.	Chinese Fan Palm	Palmae	Palm
98	*Lonicera japonica* Thunb.	Japanese honeysuckle	Caprifoliaceae	Climber
99	*Malphigia grandiflora* Jacq.	Malphigia	Malphiginaceae	Shrub
100	*Melia azadirachta* L.	Drek	Meliaceae	Tree
101	*Milletia ovalifolia* Kurz.	Milletia	Papilionaceae	Tree
102	*Morus alba* L.	Tut	Moraceae	Tree
103	*Murraya exotica* L.	Murraya	Rutaceae	Tree
104	*Murraya koengii* (L.) Spreng.	Kurry patta	Rutaceae	Tree
105	*Musa paradisiaca* L.	Banana	Musaceae	Herb
106	*Nandina domestica* Thunb.	Heavenly bamboo, Nandina	Berberidaceae	Shrub
107	*Nerium indicum* Mill.	Kaner	Apocynaceae	Shrub
108	*Nerium odorum Soland.*	Indian oleander	Apocynaceae	Shrub
109	*Ochna squarrosa* L.	Golden Champak	Ochnaceae	Tree
110	*Oncoba spinosa* Forsk.	Snuff-box tree	Flacourtiaceae	Shrub

Sl. No.	*Name of the Plant/ Botanical name*	*Common Name*	*Family*	*Habit*
111	*Roystonea regia (Kunth) O.F.Cook*	Cuban Royal palm	Palmae	Palm
112	*Oroxylum indicum* Vent.	Broken bones tree	Bignoniaceae	Tree
113	*Parkinsonia aculeate* L.	Vilayati Kikar	Caesalpiniaceae	Shrub
114	*Phoenix dactylifera* L.	Khajoor	Palmae	Palm
115	*Pinus longifolia* Roxb.	Chir Pine	Pinaceae	Tree
116	*Pinus roxburghii* Sarg.	Chir	Pinaceae	Tree
117	*Pithecellobium dulce* Benth.	Jangli Jalebi	Mimosaceae	Tree
118	*Plumeria rubra* L.	Champa	Apocynaceae	Shrub
119	*Podocarpus species*	Podocarpus	Podocarpaceae	Shrub
120	*Polyalthia longifolia* Thw	Ashoka	Annonaceae	Tree
121	*Polyalthia species*	Polyalthia	Annonaceae	Tree
122	*Porana paniculata* Roxb.	Bridal creeper	Convulvulaceae	Climber
123	*Prosopis juliflora* DC.	Vilayati Kikar	Mimosaceae	Tree
124	*Prunus persica* Batsch	Aru	Rosaceae	Tree
125	*Psidium guajava* L.	Amrud	Myrtaceae	Tree
126	*Pterospermum acerifolium* Willd.	Kanak Champa	Sterculiaceae	Tree
127	*Punica granatum L.*	Anar	Lythraceae	Tree
128	*Putranjiva roxburghii* Wall.	Putranjiva, Putijiva	Euphorbiaceae	Tree
129	*Pyrus communis* L.	European pear	Rosaceae	Tree
130	*Quisqualis indica* L.	Rangoon ki Bel	Combretaceae	Climber
131	*Rosa indica* L.	Gulab	Rosaceae	Shrub
132	*Russelia floribunda* Kunth	Firecracker plants	Plantaginaceae	Shrub
133	*Sapindus trifoliatus* L.	Reetha	Sapindaceae	Tree
134	*Sapium sebiferum* Roxb	Pippalyang	Sapindaceae	Tree
135	*Saraca indica* Auct.	Ashoka	Caesalpiniaceae	Tree
136	*Schleichera oleosa* (Lour.) Oken.	Kusum	Sapindaceae	Tree
137	*Serrisa foetida* Willd.	Yellow-rim	Rubiaceae	Shrub
138	*Spiraea corymbosa* (Raf.) Maxim.	Spiraea	Rosaceae	Shrub
139	*Sterculia alata* Roxb.	Budddha coconut	Sterculiaceae	Tree
140	*Sterlitzia reginae* Aiton	Bird of paradise	Strelitziaceae	Shrub
141	*Tabernaemontana coronaria* R. Br.	Chandni	Apocynaceae	Shrub
142	*Tecoma grandiflora Loisel.*	Trumpet creeper	Bignoniaceae	Climber shrub
143	*Tecoma stans* (L.) H.B. and K	Yellow, Trumpet bush	Bignoniaceae	Tree
144	*Tectona grandis* L.f.	Sagwan	Verbenaceae	Tree
145	*Terminalia arjuna* (Roxb.) Wight and Arn	Arjuna	Combretaceae	Tree
146	*Terminalia bellerica* Roxb.	Bahera	Combretaceae	Tree
147	*Terminalia chebula* Retz.	Harra, Harrar	Combretaceae	Tree

Sl. No.	*Name of the Plant/ Botanical name*	*Common Name*	*Family*	*Habit*
148	*Thevetia nerifolia* Juss. ex steud.	Lucky nut tree	Apocynaceae	Shrub
149	*Thunbergia erecta* (Benth.) T.Anderson	Bush Clock Vine	Acanthaceae	Shrub
150	*Tylophora asthamatica* Wight and Arn.	Jangli -pikvam	Asclepiadaceae	Climber
151	*Vitex agnus* L.	Chaste-tree	Verbenaceae	Tree
152	*Vitex negundo* L.	Sambhalu	Verbenaceae	Tree
153	*Wisteria sinensis* Sweet	Chinese wisteria	Papilionaceae	Shrub
154	*Withania somnifera* Dunal	Ashwagandha	Solanaceae	Herb
155	*Yucca gloriosa* L.	Spanish dagger	Agavaceae	Shrub
156	*Zamia sp.*	Zamia	Cycadaceae	Gymnosperm
157	*Zizyphus jujube* Mill.	Ber	Rhamnaceae	Tree

Figure 11.4: *Sterlizia reginae* Aiton.

Figure 11.5: *Juniperus chinensis* L.

Figure 11.6: *Russelia floribunda* Kunth.

Figure 11.7: *Bauhinia vahlii* Wight and Arn.

Figure 11.8: ***Ficus infectoria*** **Roxb.**

Figure 11.9: ***Casuarina equisitifolia*** **L.**

Figure 11.10: ***Oroxylum indicum*** **Vent.**

Figure 11.11: ***Oncoba spinosa*** **Forsk.**

References

Almonds E. 1993. Changing images. In: Perez J.D.R. (Ed.) Cultivating green awareness Spain: Jardin Botanico Canario- Vieray Clavijo, pp.65-68.

Balick M.J. 1986. Botanical gardens and arboreta: Future directions. New York: American Association of Botanic Gardens and Arboreta.

Bamber C. J. 1976. Plants of the Punjab. Bishen Singh Mahendra Pal Singh, Dehradun.

Chakravorty R. K. and Mukhopadhyay D.P. 1990. A Directory of botanic gardens and parks in India Calcutta: Botanical survey of India. pp.10.

Dompka V. 1996. Human population, biodiversity and protected areas: science and policy issue. American Association for the Advancement of Science, Washington, D.C.

Hooker J.D. 1885. Flora of British India. Volume 2. L. Reeve and Co., Ltd., Ashfort, Kent.

Hooker J.D. 1885. Flora of British India.Volume 3. L. Reeve and Co. Ltd., Ashfort, Kent.

Hooke J.D. 1885. Flora of British India. Volume 4. L. Reeve and Co.,Ltd., Ashfort, Kent.

Hooker J.D. 1885. Flora of British India. Volume 5. L. Reeve and Co., Ltd., Ashfort, Kent.

Maunder M. 1994. Botanic Gardens: Future Challenges and Responsibilities. *Biodiversity and Conservation* 3: 97-103.

Mittermeier R. A., Myers N., Gil P.R. and Mittermeier C.G. 1999. Hotspots: Earth's biodiversity richest and most threatened ecosystems. CEMEX, Mexico, D.F.

Nair N.C. 1978. The Flora of the Punjab Plains. BSI, Indian Botanic Garden, Howrah.

Sabnis T.S. 1986. The Flora of Punjab Plants and the associated hill regions. Scientific Publishers, Jodhpur

Sharma M. 1990. Punjab Plants Check-list. Bishen Singh Mahendra Pal Singh, Dehradun.

Sharma S. and Bir S.S.1978. Flora of Patiala: An annotated catalogue of the Wild, Naturalized and cultivated vascular plants of Patiala, Punjabi University, Patiala.

Steward J.L. 1977. Punjab Plants. Periodical Experts Book Agency, Vivek Vihar, Delhi.

Stewart J.U. and Brandis D. 1972. The Forest Flora of North-west and Central India. Bishen Singh Mahendra Pal Singh, Dehradun.

Roxburg W. 1971. Flora Indica or Description of Indian plants. Today and Tomorrow Printers and Publishers, New Delhi.

Talbot W.A. 1976. Forest Flora of the Bombay Presidency and Sind. Today and Tomorrow Printers and Publishers, New Delhi.

Thompson K. And Jones A. 1999. Human population density and prediction of local plant extinction in Britain. *Conservation Biology* 13: 185-189.

Chapter 12

Herb Garden: An Important Component of Botanic Garden for *Ex-situ* Conservation of Medicinal and Aromatic Plants

R.C. Nainwal, D. Singh, R.S. Katiyar, S. Sharma, A.K. Singh, P.K. Singh and S.K. Tewari

CSIR-National Botanical Research Institute, Rana Pratap Marg, Lucknow – 226 001, U.P.
E-mail: nainwal.rakesh@gmail.com

Abstract

Botanical gardens are ideal place to practice ex situ conservation because they have appropriate facilities and skilled horticulturists and botanists. Ex situ conservation includes not only the cultivation of plants in gardens and green houses, but also maintenance of seed, pollen or propagule samples and in vitro cell and tissue cultures. Botanic gardens also play a vital role in conservation of medicinal and aromatic plants (MAPs) since they increase the level of awareness about these along with their values. Such gardens, referred mostly as "Herb Garden" are also useful for conducting ex situ cultivation programmes and maintaining collections of these plant species. From a long time, CSIR- National Botanical Research Institute, Lucknow is actively engaged in the ex situ conservation of different types of plants and herbs, in its Botanic Garden and at Distant Research Centre, Banthara (26°42′ N, 80°49′E, 122.7m amsl). The group at Banthra is actively engaged in collection, introduction, conservation and multiplication of germplasm of MAPs, collected from various parts of country in the Herb Garden. For cultivation of MAPs in wastelands (salt affected soils), one of the propositions is searching for salt tolerant non-traditional crops, which can withstand under such stress conditions. The associated objective of the Herb Garden is searching promising medicinal and aromatic crops for sodic land and developing their agro-techniques. Various types of plants, conserved in the Herb Garden, include trees in Avicenna Avenue, small trees and shrubs in shrub plot, herbs (annual and perennial) in the garden and aquatic plants in the ponds. The Herb Garden is supported by a nursery for their propagation and processing and storage shed.

Keywords: *Botanic garden, Ex-situ conservation, Medicinal plants, Aromatic plants.*

Introduction

Plant diversity is considered as remarkable to support better regulation of various ecosystems. In nature, organisms are interlinked, and therefore, diversity of the plants supports the natural eco-system. The richness of the plants is variable among different ecological sites. The adaption capacity and tolerance capacity of the plants in natural sites are the key factors for their presence and sustainability. Production, dispersal, germination *etc.*, are key factors for development of a plant population in certain ecological areas. India is marked for its rich biodiversity due to varied climatic conditions, and is one of 12 mega biodiversity centers, including around 45,000 plant species, out of which approximately 15,000 are flowering plants. Diversity of the species is responsible for formation of complex ecosystem in nature. Out of the rich plant diversity, many of them are significantly used for varied purposes by the human beings. A group is also marked for their medicinal purposes, used for treatments of various disorders referred as medicinal plants. Some plants includes aroma variable in plant to plant species are known as aromatic plants. Association of the various medicinal and aromatic plants (MAPs) are remarkable for their use by the human beings in different localities. Around 80 per cent human beings in the world, are utilizing medicinal plants for treatment of various disorders (Kamboj, 2000).

Due to day by day increasing population of human beings, their requirements also parallel increase. For fulfilment of the basic requirements, plant diversity is directly or indirectly being interfered. One side because of over exploitation, forest fire, overburdened population, diseases, introduction of the new species in certain ecological localities *etc.*, and secondly destruction of the natural habitat, are key factors for loss of the plant species from the nature. On the basis of above key points an urgent need for the conservation of biodiversity is required globally. Conservational study on Medicinal plants were carried out by the researchers like Adhikari *et al.* (2007), Ansari (1993), Deshmukh (2010), Kasagana and Karumuri (2011), Kayombo *et al.* (2013), Raina *et al.* (2011), Rajkumar *et al.* (2011), FRLHT (2006). Mazid *et al.* (2012) focused on Medicinal Plants of Rural India: A Review of Use by Indian Folks. Study related to Herbal Garden development and use was assessed by Heywood (1983), Rao and Das (2011), Stirton (1998), Thacker (1979). Shankar and Rawat (2013) recorded Conservation and cultivation of threatened and high valued medicinal plants in North East India.

Cultivation of the MAPs is sustainable alternative for collection, propagation and conservation of the varied plants of traditional medicinal ant other important plants (Anon 2001). Herbal Garden playing a major role in *ex-situ* conservation of the MAPs of varied localities. It should be developed and managed in proper way for sustainable conservation of the plant species. Data base creation, digital herbarium preparation *etc.*, are important for circulation of the scientific knowledge on the MAPs in a wide range. It also helps for identification, further cultivation and conservation of the useful plants of medicinal values.

Herbal Garden

It is an example of *ex-situ* conservation of MAPs, introduced for their conservation and multiplication purpose. The main objectives are:

- To conserve MAPs and demonstrate their cultivation technology
- To establish a gene pool of indigenous and exotic species for conservation, research and propagation
- To popularize the cultivation and use of MAPs in the area amongst local people
- To establish MAPs resource base on sustainable basis
- To develop a centre for tourist/visitors attraction to help popularize the Indian Systems of medicine
- To raise and distribute quality MAPs seedling or its propagating material among the local farmers.

Categorization of the Plants in Herbal Garden

Categorization of the MAPs introduced in herbal garden should be on the basis of their utility as well as their habits as herb, shrub, tree, climber, creeper, underground MAPs, aromatic plants.

- Medium height trees and shrubs should be planted in certain distance.
- Short herbaceous/climber and underground plants should be planted in between medium height trees and shrubs, to provide support and shade in high temperature range during summer season.
- Based on importance and their population strength classification can be done.
- More protection should be provided to RET and endangered MAPs in herbal garden.
- All the introduced MAPs should be protected from all biotic and abiotic factors which directly or indirectly affecting the plant life pattern.

Medicinal Plants Conservation Areas (MPCA) are the forest patches related for conservation as well as for supporting propagation of the prioritized, RET, endangered and valuable MAPs in different ecological areas.

The Conservation of MAPs is a challenging task in current scenario due to following reasons:

1. Over exploitation
2. Increasing Human population load
3. Destruction of natural habitat
4. Reduction of forest areas
5. Introduction of the new species
6. Environmental pollution
7. Forest fire
8. Less awareness about the valuation of the Biodiversity *etc.*

Establishment of Herbal Garden

It includes several following requirements

Site Development

It is the major step for development of any garden, which includes better soil quality, availability of the facilities needed for the plants, transport facility for supply of the materials required for its development as well as plant supply for their further conservation.

Preparation of the Site for Garden

Site preparation is considered as the second most important step for garden development. Preparation of the field accomplished by ploughing, harrowing and slope formation. After of these practices field require preparation of the well-designed beds for every single species of the MAPs.

Fencing

The selected area of the garden need to be properly protected through fencing around the marked areas for protection of the introduced spp. of MAPs for their *ex-situ* conservation.

Water Arrangement for Irrigation Facilities

Better water facilities are urgent need in garden. Water availability should be made for each beds/plants by providing irrigation water using small canals or by other modes.

Cultivation/Sowing of MAPs

Collected/Selected plants/plant parts should be grown/sown in the prepared beds following their recommended distance managing from plant to plant and row to row. Modified vegetative plant parts like bulb, tuber, rhizome, corm *etc.*, and the seeds are directly grown/sown in the prepared beds of the garden, however new plantlets development was done using the same parts of the plants in poly bags. After maturation of the small plants are carefully transferred in to the selected sites followed by supply of the facilities as per need of the plants.

Monitoring/Supervision

Cultivated/grown plants were daily monitored to examine the need of the plants and every possible effort was made to support the better plant growth and development. Initially newly developing plants were protected from high intensity of sunlight by covering paper/bamboo cartoons during day time and is opened at evening following till the establishment of new individuals in the field.

Management Strategies

Water Management

Water requirement of any plant is vary depends upon the species, stage and the location. Water is important for successful physiological activities in the plants that support better growth and development. A moderate range of water is beneficial for the plants and is variable among the varied plant species. Excess water is responsible for water logging near the roots of the plants adversely affecting absorption of water, nutrients. It also disturbs soil aeration and finally being reason for death/decay of the plants. Deficiency of water also adversely affects the plant life showing their symptoms as wilting of aerial parts of the plant. Staying in such condition for long time leads to death of the plants. So water should be provided as per need of the plants grown for *ex-situ* purpose in herbal garden.

Weed Management

Weeds are undesirable/unwanted populations of the plants belonging to different habits as well as families. They compete for light, space, nutrients *etc* with the grown plants in the field so they are needed to remove from the field without damage of the cultivated plants. Weeds removal should be made before their flowering to avoid their numerous seed production.

Nutrient Management

Nutrients play very important role in plants for regulation of the various physiological activities and formation of compounds/parts. Macro and micronutrients should be supplied in the field as per need of the cultivated plant spp.

Disease Management

Plant diseases are abnormal conditions in the plants, which can be observed by appeared symptoms by daily monitoring. Proper diseases management helps to keep the plants healthy that further support better life pattern and leads much production/output.

Insects/Pests/Rodents Management

These are very harmful for the plants. However, among the rich diversity of the insects some are remarked as important, for their role in participation for pollination in the flowers, needed for development of the seeds in plants. Harmful insects, pests and rodents should be keep away by using suitable chemicals and removing alternate host plants/weeds from the field. By following above parameters/suitable methods, a herbal garden has been developed for *ex–situ* conservation of the MAPs, at Distant Research Centre, Banthara of CSIR-National Botanical Research Institute, Lucknow. Where many important, rare, endangered species of the MAPs were introduced from different parts of the country, for their long term presence and utilization purpose (Tables 12.1–12.4). It is playing a significant role in North India aimed for *ex-situ* conservation of the MAPs.

Table 12.1: List of Conserved Herbs at Distant Research Centre, Banthra

Sl.No.	Scientific Name	Common Name	Family
1	*Abelmoschus mosdhatus* Medic	Mushkdana, Ambrette	Malvaceae
2	*Abutilon indicum* L.	Kanghi Kankati	Malvaceae
3	*Achyranthes aspera* L	Latjira, Apamarg	Amaranthaceae
4	*Acorus calamus* L.	Bach, Vacha	Araceae
5	*Allium cepa* L.	Piyaz, Palandu	Liliaceae
6	*Allium sativum* L.	Lahsun, Sashuna	Liliaceae
7	*Allium schoenoprasum* L.	Chive	Liliaceae
8	*Aloe vera* (L.) Burm. f.	Ghritkumari	Liliaceae
9	*Alipinia galanga* (L.)Willd.	Kulanjan	Zingiberaceae
10	*Althaea rosea L. Cav. syn. Alcearosea L.*	Khatmi	Malvaceae
11	*Amaranthus hypochondriacus* L.	Ramdana Atrilal	Amaranthaceae
12	*Ammi majus* L.	Atrilal	Apiaceae
13	*Ammi visnaga*	toothpick-plant, bisnaga	Apiaceae
14	*Amorphophallus paeoniifolius* (Dennst.) Nicolson	Zaminkand, Suran	Araceae
15	*Andrographis paniculata* (Burm.f.)Wall. ex. Nees	Kalmegh, Harachiraita	Acanthaceae
16	*Anethum sowa* Roxb. ex. DC. syn. *Anethum graveolens* L.	Sowa, Soya.	Apiaceae
17	*Apium graveolens* L.	Ajmod, Ajmoda	Apiaceae
18	Artemisia sieversiana Ehrh	Dauna	Asteraceae
19	*Asparagus racemosus* Willd.	Satawar, Shatawari	Liliaceae
20	*Avena sativa* L.	Jau Birahna, Jayee	Poaceae
21	*Bacopa mdnnieri* L	Jalneem, Brahmi	Scrophulariaceae
22	*Beta vulgaris* L.	Chukander	Chenopodiacea.
23	*Boerhavia diffusa* L.	Punarnava	Nyctaginaceae
24	*Brassica compestris* L	Sarson	Brassicaceae
25	*Calotropis gigantea* (L.) R.Br.	Madar, Aak, Arka	Asclepiadaceae
26	*Calotropis procera* (Ait.) R.Br.	Madar, Aak	Asclepiadaceae
27	*Cannabis sativa* L.	Bhang, Ganja	Cannabinaceae
28	*Carthamus tinctorius* L.	Qurtum, Kusum	Asteraceae
29	*Cassia angustifolia* Vahl.syn. Cassia senna L.	Senna, Sanamukhi	Caesalpiniaceae
30	*Cassia tora* L.	Chakunda, Chakramarda	Caesalpiniaceae
31	*Catharanthus roseus* (L)G.Don	Sadabahar	Apocynaceae

Sl.No.	Scientific Name	Common Name	Family
32	*Centella asiatica* (L.) Urban.	Mandukparni, Brahmi	Apiaceae
33	*Centratherum anthelminticum* (Willd.) Kuntze	Kalijiri, Somraj	Asteraceae
34	*Chamomilla recutita* Syn. *Matricaria chamomilla* L.	Baboonah	Asteraceae
35	*Chenopodium ambrosioides* L	American Wormseed	Chenopodiaceae
36	*Cichorium intybus* L	Kasni	Asteraceae
37	*Cissus quadrangularis* L.	Hadjod	Vitaceae
38	*Clerodendrum indicum* (L.) Ktze	Bharangi	Verbenaceae
39	*Clerodendrum infortunatum* L.	Bhant	Verbenaceae
40	*Coleus amboinicus* Lour	Paatherchur	Lamiaceae
41	*Coriandrum sativum* L.	Dhania	Apiaceae
42	*Costus speciosus* (Koen). Sm.	Qust Talakh, Keu	Zingiberaceae
43	*Crinum asiaticum* L.	Sudarsan, Nagdamni	Amaryliidaceae
44	*Cuminum cyminum* L.	Zeera	Apiaceae
45	Curcuma amada	Ama haldi	Zingiberaceae
46	*Curcuma longa* L.	Halid, Haridra	Zingiberaceae
47	*Cymbopogon citratus* (DC) Stapf	Lemongrass	Poaceae
48	*C.flexuosus* (Nees ex Steud.)	Lemongrass	Poaceae
49	*C. martinii* (Roxb.) Watson	Rosha, Palmarosa	Poaceae
50	*C. nardus* L.	Citronella	Poaceae
51	*Cyperus scariosus* R.Br,	Nagar Motha	Cyperaceae
52	*Datura metel* L.	Dhatura, Taatur	Solanaceae
53	*Desmedium gangeticum* D	Sarivan	Leguminoceae
54	*Eruca sativa* Mill	Jirjir, Taramira	Brassicaceae
55	*Euphorbia hirta* L.	Bari dudhi	Euphorbiaceae
56	*E. thymifolia* L.	Chhoti dudhi	Euphorbiaceae
57	*Evolvulus alsinoides*	Shankpushpi	Convolvulaceae
58	*Foeniculum vulgare* Mill.	Badi Saunf	Apiaceae
59	*Giycyrrhiza glabra* L.	Mulothi, Asnlussoos	Fabncono
60	*Hibiscus mutabilis* L.	Gulajaib	Malvaceae
61	*Hordeum vulgare* L	Jau	Poaceae
62	*Hyoscymus niger* L	Ajwain, Khorasani	Soianaceae
63	*Indigofera tinctoria*	Neel	Fabaceae
64	*Ipomoea batatas* Lam.	Shakerkand	Convolvulaceae
65	*Lactuca sativa* L.	Kahu	Asteraceae
66	*Lepidium sativum* L	Halim, Chansur	Brassicaceae
67	*Leucas aspera Spreng.*	Goma, Dronpushpi	Lamiaceae
68	*Linum usitatissimum* L.	Alsr, Atsi	Linaceae
69	*Malva rotundifolia* L.	Khubbazi	Malvaceae

Sl.No.	Scientific Name	Common Name	Family
70	*Matricaria chamomile* L.	Babunah	Asteraceae
71	*Melilotus indica* L.	Banmethi	Leguminosae
72	*Mentha arvensis* L.	Podina	Laminaceae
73	*Mentha citrata* L.	Podina	Lamiaceae
74	*Mentha piperita* L	Podina	Lamiaceae
75	*Mentha viridis* L.	Podina	Lamiaceae
76	*Mimosa pudica* L.	Lajwanti	Fabaceae
77	*Mirabilis jalapa* L.	Gulabbas	Nyctaginaceae
78	*Nelumbo nucifera* Gaertn	Kanwal, Kamal	Nymphaceae
79	*Nymphaea alba* L.	Nilofar, Kumudni	Nymphaeaceae
80	*Ocimum americanum* L.	Kalitulsi	Lamiaceae
81	*Ocimum basilicum* L.	Ban Tulsi	Limiaceae
82	*Ocimum sanctum* L.	Tulsi, Manjari	Lamiaceae
83	*Papaver somniferum* L.	Posta, Khashkhash	Papaveraceae
84	*Pedalium murex* L.	Baragokhru	Pedaliaceae
85	*Phyllanthus niruri* L.	Bhui amla	Euphorbiaceae
86	*Plantago ovata Forsk.*	Isabgol	Plantaginaceae
87	*Pluchea lanceolata* Oliver and Hiern.	Rasna	Asteraceae
88	*Plumbago zeylanica* L.	Chitcak	Plumbaginaceae
89	*Portulaca oleracea* L.	Kulfa, Lonika	Portulacaceae
90	*Psoralea corylifolia* L.	Babchi, Bakuchi	LeguTiinosae
91	*Raphanus sativus* L.	Muli, Mulaka	Brassicaceae
92	*Rosa damescena Mill*	Fasali Gulab	Rosaceae
93	*Sesamum indicum* L.	Til	Pedaliaceae
94	*Sida acuta Baurm.*	Bala	Malvaceae
95	*Sida cordifolia* L.	Brela	Malvaceae
96	*Sida rhombifolio* L.	Atibala, Swet bala	Malvaceae
97	*Solanum nigrum* L.	Makoi, Kakamachi	Solanaceae
98	*Solanum xanthocarpum* Schard. and Wendle	Kateli Kantakari	Solanaceae
99	*Sphaeranthus indicus* L.	Mundi	Asteraceae
100	*Tagetes erecta* L.	Genda, Marigold	Asteraceae
101	*Trachyspermum ammi (L.)* Spr.	Ajwain	Apiaceae
102	*Trianthema pentandra* L.	Biskhapra	Ficoidaceae
103	*Tribulus terrestris* L.	Chhota Gokhru	Zygophyllaceae
104	*Trigonella foenum-graecum* L.	Methi, Chandrika	Fabaceae
105	*Uraria picta* Desv.	Pithvan	Fabaceae
106	*Vernonia cinerea* Less.	Sahadevi	Asteraceae
107	*Vetiveria zizanioides (L)* Nash	Khas	Poaceae
108	*Vicia faba* L.	Bakla	Fabaceae

Sl.No.	*Scientific Name*	*Common Name*	*Family*
109	*Vinca rosea* L.	Sadabahar	Apocynaceae
110	*Wedelia calendulacea* Less.	Pila bhangra	Asteraceae
111	*Withania somnifera (L)* Dunal	Asgand, Ashwagandha	Solanaceae
112	*Zingiber officinale* Rose.	Adrak	Zingiberaceae
113	*Zingiber zermubet (L.)* Roscoe ex Sm.	Narkachur	Zingiberaceae

Table 12.2: List of Conserved Shrubs at Distant Research Centre, Banthra

Sl.No.	*Scientific Name*	*Common Name*	*Family*
1	*Adhatoda vasica* Nees.	Adusa, Vasaka	Acanthaceae
2	*Annona squamosa* L.	Sharifa, Sitaphal	Annonaceae
3	*Barleria prionitis* L.	Piyabasa.Katsareya	Acanthaceae
4	*Bixa orellana* L.	Latkan.Sinduri	Bixaceae
5	*Bluemea lacera* D.C.	Kukraundha	Asteraceae
6	*Caesalpinia crista* L.	Karanj, Katkaianj	Caesalpiiniaceae
7	*Carissa carandas* L	Karaunda, Karmardak	Apocynaceae
8	*Cassia alata* L.	Dadmurdan	Caesalpiniaceae
9	*Cassia occidentalis* L.	Kasaundi, Kasmard	Caesalpiniaceae
10	*Celastrus paniculatus* Willd.	Malkangnj, Jyotishmati	Celastraceae
11	*Cestrum diurnum*	Din ka raja	Solanaceae
12	*Citrus aurantifolia* (Christm.) Swingle	Kagji nibu	Ruta-ceae
13	*Citrus medica* L.	Turang, Matulung	Rutaceae
14	*Citrus reticulata* Blanco.	Narangi	Rutaceae
15	*Citrus sinensis* (L.) Osbeck	Mausambi	Rutaceae
16	*Ficus carica* L	Anjeer	Moraceae
17	*Glycosmis pentaphylla* Corr.	Ban-nibu	Rutaceae
18	*Grewia asiatica* L.	Falsa, Parusha	Tiliaceae
19	*Gymnema sylvestre* R.Br.	Gurmar	Asclepiadaceae
20	*Hemidesmus indicus* R.Br.	Anantmool	Asclepiadaceae
21	*Hibiscus rosa-sinensis* L.	Gurhal	Malvaceae
22	*Hibiscus tiliaceus* L.	Bellipata	Malvaceae
23	*Jasminum sambac* (L) Ait.	Motia, Bela, Mogra	Oleaceae
24	*Jatropha curcas* L.	Bagbherenda	Euphorbiaceae
25	*Lagerstroemia indica* L.	Sawani	Lythraceae
26	*Lantana camara* L.	Ghaneri	Verbenaceae
27	*Lawsonia inermis* L.	Menhdi	Lythraceae
28	*Leucaena glauca* Benth.	Bilayeti babul	Leguminosae
29	*Murraya koenigii* Spreng.	Katnim	Rutaceae
30	*Murraya exotica* L.	Marchula, Kamini	Rutaceae
31	*Musa paradisiaca* L.	Kela	Musaceae

Sl.No.	Scientific Name	Common Name	Family
32	*Nerium indicum* Mill	Kaner Lai	Apocynaceae
33	*Nyctanthes arbortristis* L.	Harsingar	Oleaceae
34	*Pandanus fascicularis* Lamk	Ketki	Pandanaceae
35	*Punica granatum* L	Gulnar	Puniaceae
36	*Rauvolfia serpentina* Benth.	Sarpagandha	Apocynaceae
37	*Smilax zeylanica* L.	Jangli ausbhah	Liliaceae
38	*Thevetia peruviana* (Pers.) Schum.	Pila Kaner	Apocynaceae
39	*Vitex negundo* L.	Sambhalqo	Verbenaceae

Table 12.3: List of Conserved Trees at Distant Research Centre, Banthra

Sl.No.	Scientific Name	Common Name	Family
1	*Acacia catechu* Willd.	Katha, Khair	Fabaceae
2	*Acacia nilotica* Willd.	Kikar, Babul	Fabaceae
3	*Adansonia digitata* L.	Gorakshi.Gorakhimii	Bgmbacaceae
4	*Aegle marmelos* (L) Corr.	Bilva, Bael	Rutaceae
5	*Alangium salvifoilum* (L) Wang	Ankol	Alangiaceae
6	*Albizia lebbeck* (L) Willd	Kala siris	Fabaceae
7	*Albizia procera* (Roxb) Benth	Safed siris	Fabaceae
8	*Alstonia scholaris* (L.) R.Br	Saptaparna.Chhatium	Apocynaceae
9	*Amoora rohituka* W and A	Rohitaka	Meliaceae
10	*Artocarpus lakoocha* Roxb.	Barhal	Moraceae
11	*Azadirachta indica*	Neem, Nimb	Meliaceae
12	*Bauhinia variegata* L.	Kachnar, Kanchanara	Fabaceae
13	*Butea monosperma* (Lam.)Taub.	Dhak, Palash, Tesu	Fabaceae
14	*Cassia fistula* L.	Amaltas	Caesalpiniaceae
15	*Cordia dichotoma* Forst.f.	Sipistan, Lasora	Boraginaceae
16	*Crataeva nurvala* Buch. Ham.	Barun	Capparidaceae
17	*Dillenia indica* L.	Chalta	Dilleniceae
18	*Diospyros montana* Roxb.	Bistendu, Aabnoos	Ebenaceae
19	*Diospyros peregrina* Gurke	Tendu, Aabnoos	Ebenaceae
20	*Eclipta alba* Hassk.	Bhangra bhringraj	Asteraceae
21	*Emblica officinalis* Gaertn	Amla	Euphorbiaceae
22	*Erythrina indica* L.	Mura	Fabaceae
23	*Eucalyptus citriodora* Hook	Eucalyptus	Myrtaceae
24	*E. globulus* Labill.	Blue gum	Myrtaceae
25	*Ferunia elephantopus*	Kaitha	Rutaceae
26	*Ficus bengaiensis* L.	Bargad	Moraceae
27	*Ficus racemosa* L.	Gular, Udumbar	Moraceae
28	*Ficus religiosa* L.	Pipal, Pippal	Moraceae

Sl.No.	Scientific Name	Common Name	Family
29	*Ficus retusa* L.	Kamrup	Moraceae
30	*Holarrhena antldysentrica* (Roth.) A.DC	Inderjau, Kurchi	Apocynaceae
31	*Holoptelea integrifolia* (Roxb.) Planch.	Papri	Ulmaceae
32	*Litsea chinensis* Lam.	Maida lakri	Lauraceae
33	*Mangifera indica* L.	Aam	Anacardiaceae
34	*Melia azedarach* L.	Bakain, Mahanimba	Meliaceae
35	*Michelia champaca* L.	Champa	Magnoliaceae
36	*Moringa oliefera* Lam.	Sahjan, Sobhanjana	Maringaceae
37	*Morus alba* L.	Shahtut, Tut	Moracoao
38	*Polyalthia longifolia Benth. and Hook.f.*	Ashok	Annonaceae
39	*Pongamia pinnata* (L.) Merr.	Karanj, Karanja	Fabaceae
40	*Psidium guajava* L.	Amrud	Myrtaceae
41	*Pterocarpus marsupium* Roxb.	Bijasal, Bajasar	Leguminosae
42	*Pterospermum acerifolium* Willd	Kanak champa	Sterculiacoae
43	*Pyrus communis* L.	Nashpati	Rosaceae
44	*Bombax ceiba* L.	Semal, Shalmali	Bombacaceae
45	*Santalum album* L.	Shwet chandan	Santalaceae
46	*Sapindus trifoliatus* L.	Ritha, Phenila	Sapindaceae
47	*Saraca asoca* L.	Ashok	Caesalpiniaceae
48	*Shorea robusta* Gaertn.	Sal	Dipterocarpaceae
49	*Strychnos nuxvomica* L.	Kuchla.Dirghpatra	Loganiaceae
50	*Syzigium cumini* (L.) Skeels	Jamum, Jambu	Myrtaceae
51	*Tamarindus indica* (L.) All.	Imli	Caesalpiniaceae
52	*Tootona grandic* L.f.	Sagaun	Verbenaceae
53	*Terminalia* arnijunn W and A,	Arjun	Combietnceae
54	*Terminalia belerica* Roxb.	Bahera	Combretaceae
55	*Terminalia chebula* Retz.	Halela, Har	Combretaceae
56	*Thespesia populnea* (L) Sol. ex Correa	Paraspipal, Porus	Malvaceae
57	*Zizyphus jujuba* Mill.	Unnab,Ber	Rhamnaceae
58	*Zizyphus mauriiiana* Lam.	Ber	Rhamnaceae

Table 12.4: List of Conserved Climber/Creeper at Distant Research Centre, Banthra

Sl.No.	Scientific Name	Common Name	Family
1	*Abrus precatorius* L.	Gongchi, Ratti	Fabaceae
2	*Acacia concinna* D.C.	Shikakai	Fabaceae
3	*Cardiospermum halicacabum* L	Kanphuti.Karnasfota	Sapindaceae
4	*Clitoria ternatea* L.	Aparajit	Leguminoceae
5	*Cuscuta reflexa* Roxb.	Amarbel	Convolvulaceae
6	*Dioscorea bulbifera* L.	Ratalu	Dioscoreaceae

Sl.No.	Scientific Name	Common Name	Family
7	*Dioscorea hispida* Dennst.	Karukandu	Dioscoreaceae
8	*Dolichos lablab* L.	Sem	Fabaceae
9	*Jasminum officinale* L.	Yasmin, Chameli	Oleaceae
10	*Kalanchoe pinnata* Pers.	Zakhm-e-hayat, Ghau Patta	Crassulaceae
11	*Mucuna nivea* (Roxb.) DC.	Kewanch	Fabaceae
12	*Mucuna pruriens* (L.) DC.	Kewanch	Fabaceae
13	*Mucuna utilis*	Kewanch	Fabaceae
14	*Operculina turpethum* (L.) Silva Manso	Turbud, Nisoth	Convolvuiaceae
15	*Piper beitle* L.	Paan	Piperaceae
16	*Piper longum* L.	Pipalli	Piperaceae
17	*Temmnus inbialis* Spreng,	Mushaparni	Fnhnnoao
18	*Tinospora cordifolia* (Willd.) Hook.f. and Th.	Giloe, Gurucha	Menispermaceae
19	*Tylphora indica* (Burm.f.) Merr.	Antamul	Asclepiadaceae
20	*Vitis venifera* L.	Angoor, Draksha	Vitaceae
21	*Ziziphus* nummulaiia W and A	Jharber, Bhubadari	Rhamnaceae

Conclusion

Finally it is concluded that the MAPs are of great significance due to presence of certain chemical compounds/aroma in their plant body which are variable among the plants and also variable in different ages of the same plants. Due to natural and manmade reasons these group of the plants are coming under the boundary of endangerment. Therefore, these plants spp should be conserved by all the possible strategy which provides better chance of propagation as well as protecting them to move near the line of endangerment.

References

Adhikari B.S., Babu M.M., Saklani P.L. and Rawat G. S.2007.Distribution, Use pattern and prospects for Conservation of Medicinal plants in Uttaranchal State, India. *J. of Mountain Science,* 4(2): 155-180.

Anon., 2001.Cultivation of Medicinal plants. National Medicinal Plants Board. Deptt. of Ayush. Ministry of Health and a Family welfare, Govt. of India.

Ansari A.A., 1993. Threatened medicinal plants from Madhauli Forests of Gorakhpur. *Journal of Economic and Taxonomic Botany*17: (10) 241.

Deshmukh B. S. 2010. *Ex-situ* conservation studies on ethno-medicinal, rare, endemic plant species from Western Ghats of Maharashtra. 1(2): 1- 6.

FRLHT 2006. Conservation and adaptive management of Medicinal plants- A participatory model: Medicinal plants Conservation Areas and Medicinal plants Development Areas. FRLHT, Banglore, India, pp 58.

Heywood V.H. 1983. Botanic Gardens and Taxonomy –Their economic role. Bull. Bot. Surv. India. Vol 25 pp 134 – 147.

Kamboj V.P. 2000."Herbal Medicine-some comments", Curr.Sci., 78(1), pp. 35 - 39.

Kasagana V.N. and Karumuri S.S. 2011.Conservation of medicinal plants (Past, Present and future trends). *Journal of Pharmaceutical Science and research*. 3(8): 1378-1386.

Kayombo E.J., Mahunnah R.L.A. and Uiso F.C. 2013. Prospects and challenges of Medicinal plants conservation and Traditional medicine in Tanzania. *Agro technology*, 1(3): 2-8.

Mazid M., Khan T.A. and Mohammad F. 2012. Medicinal Plants of Rural India: A Review of Use by Indian Folks. *Indo Global Journal of Pharmaceutical Sciences*, 2(3): 286-304.

Raina, R., Chand R. and Sharma, Y.P. 2011. Conservation strategies of some Medicinal Plants. *International Journal of Medicinal and Aromatic plants*, 1(3): 342-347.

Rajkumar M.H., Sringeswara A.N. and Rajanna M.D. 2011. *Ex-situ* conservation of Medicinal plants at university of agricultural sciences, Bangalore, Karnataka. *Recent Research in Science and Technology*, 3 (4): 21-27

Rao N.S. and Das S.K. 2011. Herbal Gardens of India: A statistical analysis report. *African Journal of Biotechnology*. 10(31): 5861-5868.

Shankar R. and Rawat M.S. 2013.Conservation and cultivation of threatened and high valued medicinal plants in North East India. *International Journal of Biodiversity and Conservation*, 5 (9): 584 – 591.

Stirton, C. 1998. Education for sustainability – A Garden for a sustainable future. Root. (Dcc) pp 38 41.

Thacker, C. 1979. The history of Gardens: Beckenham: CroomHelm Publication Ltd.

Chapter 13

Conservation and Evaluation of *Canna* Cultivars for Cultivation on Partially Sodic Waste Land of Northern Plains of Uttar Pradesh

Devendra Singh, R.C. Nainwal, R.S. Katiyar and S.K. Tewari

CSIR-National Botanical Research Institute, Lucknow, Uttar Pradesh
E-mail: singhdrdevendra@gmail.com

Abstract

Scientific research in the Botanic Gardens is based on purely taxonomic research through the relevance of technical applications to conservation efforts for all plants groups, and their pragmatic applications in the floriculture, herbal and medicinal plant industries. One of the most important functions of the Botanic Gardens is the ex-situ conservation of the flora. Ex situ conservation involves the collecting, handling and management (including research) of germplasm, its storage, regeneration, characterization/evaluation, documentation and dissemination to users. Botanic Gardens conduct research on the cultivation requirements, the reproductive biology and the propagation of individual plants. Such information is essential to be able to reintroduce the plants back into the wild and to provide material for restoring and rehabilitating natural habitats. In the present era, floriculture is gaining much importance. The flowering industry has become a good source of income. Besides it also gives pleasure and happiness to human beings. A number of flowers are used for decorative purposes in homes. Canna indica is an ornamental plant having more aesthetic value. Most of the commercial cultivars of Canna are very attractive. Canna is the only genus in the family Cannaceae. It's richly varied colour and along vase life, are the reasons for its ever increasing demand. In one of the experiment conducted at Distant Research Centres of CSIR-NBRI, five Canna varieties were examined to screen the best suited cultivar for partially sodic soil of northern plains of Uttar Pradesh having pH of around 8.5. The varieties included Cattleya, Pink Sunrise, Flaccida, Black Night and Bengal Tiger. Significant variations have been seen in terms of their growth performance like plant height, number of leaves, tillers and flowers per plant. Significantly higher plant height was recorded in Black night throughout the year. This was followed by Bengal tiger and Cattleya, which were at par with Pink sunrise. However, number of tillers per plant was significantly higher in later stage of growth at 300 DAP (days of planting).

Keywords: *Canna, Ex situ, Germplasm, Sodic soil.*

INTRODUCTION

Long before the term "biodiversity" was used, botanical gardens carried out activities that are now associated with biodiversity. They took part in describing new species and studies about them to discover potential uses in industry, horticulture or for research. Gardens also conserved species of rare wild plants (or *ex situ* conservation, meaning outside of the species' natural habitat).One of the most important roles of conservation is environmental education. Each year, more than 150 million people visit gardens all over the world and have the chance to get in touch with nature. Botanical gardens are a unique environment to raise public awareness and help people understand the importance of biodiversity, educate people about the threats it currently faces and make them realize that nature conservation is everyone's job. This is why it is so important for gardens to maintain interpretation programs, host school groups and present exhibitions. *Ex situ* conservation (growing wild plants outside their natural environment) has many advantages, but should not be seen as an objective in and of itself. It is one element of a comprehensive strategy to conserve species in their environment. *Ex situ* conservation helps to attain this objective by providing material to reintroduce plants into degraded areas or to reinforce existing populations.

It also helps to remove wild populations from the pressure of scientists, horticulturists or collectors. The presence of a rare species in a botanical garden makes it available for scientific research, education and possible horticultural or commercial exploitation without affecting wild populations. *Ex situ* conservation can also serve as an "insurance policy" for endangered species by creating a protected reserve of especially vulnerable native species or populations. It can even be the only solution if the natural habitat has been destroyed or if a species disappears. Botanical gardens are ideal places to practice *ex situ* conservation because they have appropriate facilities and skilled horticulturists and botanists. *Ex situ* conservation includes not only the cultivation of plants in gardens and greenhouses, but also maintenance of seed, pollen or propagule samples and in vitro cell and tissue cultures.

Land, a non-renewable resource, is central to all soil cannot be recovered at all. Over the years, the country's countries like India due to high population rate there is landmass has suffered from different types of extreme pressure on crop lands and simultaneously in degradations. The problems of soil alkalinity, salinity and poor quality waters are likely to increase in the foreseeable future mainly due to expansion in irrigated area, intensive use of natural resources resulting in second generation problems and climate change which is no longer a fashion statement but a reality. Our tentative estimate indicates that area under salt affected soils in India may increase to 16.2 million ha by 2050. The problems of water quality especially polluted waters are also likely to increase exponentially due to increasing urbanization and industrialization. Management of these degraded land and water resources is likely to pose formidable problems because many new pollutants are emerging. Due to poor physical properties, excessive exchangeable sodium and high pH, most of the alkali lands support a poor vegetative cover. Many of the floriculture crops are in great demand for both internal requirements and export. As most of these crops are non-conventional in nature, they are not preferred on fertile lands which are used for food crops. The marginal lands, especially those affected by salinity or sodicity problems where profitable returns are not feasible from agricultural crops, could be successfully utilized for the cultivation of floricultural crops with marginal inputs. The marginal lands, especially those affected by salinity, sodicity and waterlogging problems could be successfully utilized for the cultivation of floriculture crop like *Canna* crop with marginal inputs.

Canna

Cannas are very important, with easy propagation, wide adaptability, indigenous, vibrant colour, rich diversity and perennial flowering. These grow abundantly in the humid tropical regions throughout the world. They are very hardy and thus grown easily and successfully. The flowers have many shades of colour, appear throughout the year and make a wonderful display of colour which can hardly be surpassed by any other perennial plant. Cannas are frequently grown in private gardens but more extensively in the public gardens. As the flowers do not last long as cut flowers and are very common garden plants, visitors do not show much interest to pluck the flowers or damage the plants. Cannas alone with their various shades of colour can make a garden quite attractive.

The genus *Canna* has about 50 species, native of Tropical America and Asia. In India improvement of *Canna* has chiefly been achieved through hybridization. A large number of varieties have been developed as a result of hybridization and selection and these varieties are now commonly grown in the gardens. Depending on the size and shape of the flowers, *Canna* has been classified as follows:

1. **Alipore Hybrids:** The selections are the results of 45 years of hybridization and are a great improvement on the crozy types from which they are derived. The size of the flowers markedly increased with a wide range of colour.
2. **Bouquet:** In this class an ideal variety 'cupid' has flowers on closely branched spikes. The plants of this variety are dwarf.
3. **Candleabra:** This is a distinct break. The main flower stalk branches and as many as 8-12 spikes are produced instead or two or three.

4. **Crozy and gladiolus flower**: These are improved hybrids raised by Anne in about 1850 and Vilmorin in 1880.
5. **Dreadnaught:** A great advance on the ordinary crozy or flowered *Canna* both as regards the individual flowers and bunches.
6. **Dwarf:** This type includes varieties which does not exceed the height of 70-80 cm at any time of the year and are very effective for breeding purpose.
7. **Giant or Orchid flower:** Originated in Italy and was very popular for many years. The large blooms of silky appearance resemble the Flag Iris, but not very hardy.
8. **Miniature:** This is small flowering type derived by crossing a society's dwarf hybrid with *Canna indica*. The spikes are neat and compact.

Cannas can also be divided into several classes according to the colour of the flowers:

1. **Selfs**: without spots or margin, one colour only.
2. **Spotted**: usually a shade of red on cream or yellow ground or red spots on orange of red ground.
3. Striped red on a cream or a yellow ground.
4. Margin yellow.
5. Margined with a darker shade than the ground colour.
6. Flaked red or orange on a paler ground.
7. Splashed orange on a deeper ground.

The shades of colour cover a wide range from creamy white through yellow, orange and pink to an intense maroon red. There is no blue or purple colour in *Canna*. There are also few varieties with dark leaves usually reddish purple, also producing dark red flowers. In one variety, however, the leaves are striped yellow.

Cultivation

The rhizomes are collected in the month of May and then planted to get the full benefit of the Monsoon and a strong clump forms before the winter. The soil is prepared by digging at least 50 cm deep, breaking the clods and mixing in fresh stable manure at the rate of 100 kg to a bed of 10 sq. meter. The rhizomes are buried 3 cm below the surface of the soil and thoroughly flooded. If the weather is hot and dry, shade is provided for a few days. Planting is done at a distance of 30-40 cm between the rows and the plant. Within six weeks after planting first flower spikes appears which should be removed to encourage better growth. The bed should be kept free from weeds by frequent weeding. After the rain ceases and the soil thoroughly dries, it is loosened and allowed to dry. Digging and watering of the bed are necessary at least once a month in the winter and more frequent flooding is recommended to keep the soil moist. After the first flush of flowering is finished in December, leaf-mould or cow dung manure should be applied to the beds to ensure good flowering in the second flush from April to June. *Cannas* can also be grown in 30-35 cm pot and the compost should consist of 2 parts stable manure and 1 part garden soil. The potted plants should be replanted every 6-9 months as the roots become pot bound.

NBRI Contribution on *Canna*

For improvement of the poor cut flower quality of *Canna*, CSIR-NBRI has started a major programme on improvement of *Canna* as promising floriculture crop by improving through modern biotechnology and classical breeding approaches by different groups with these objectives:

1. Molecular basis of flower colour variability in *Canna*.
2. Enhancing flower vase life in *Canna*.
3. Marker-assisted selection to improve flower traits in *Canna* Germination of *Canna* seeds: A basic problem.
4. Elucidating role of micro RNA in flower development of selected contrasting genotypes of *Canna*.
5. Regeneration and Genetic Transformation of *Canna* spp.
6. Identification and molecular characterization of insect resistant *Canna* germplasm.
7. Detection and identification of viruses occurring on *Canna* spp. and development of virus-free plants.

Major Achievements

Conservation in CSIR-NBRI Botanic Garden

The CSIR-NBRI Botanic Garden has following 9 species and also having 50 varieties of *Canna*.

1. *Canna argentina*
2. *Canna latifolia*
3. *Canna edulis*
4. *Canna flaccida*
5. *Canna generalis*
6. *Canna indica*
7. *Canna gigantia*
8. *Canna brasiliensis*
9. *Canna warscewiczii*

Varietal Improvement of *Canna*

In 2009, the new cultivar *Canna generalis* 'Agnisikha' is a mutant of *Canna generalis* 'Lucifer'. Rhizomes were treated with different doses of gamma rays (Cobalt 60) and planted in the beds. Mutated plant was identified by change of flower colour, isolated, multiplied vegetatively, evaluated in the successive generations and found distinct, uniform and stable mutant (Roy, 2004). This breeding work including irradiation of the rhizomes, field trials were carried out in the Botanic Garden.

Morpho-agronomic Characteristics

A. Vegetative

Stature – Medium (0.90 – 1.20 m)

Stem – Stout, 2.5 – 3.5 cm in diameter with moderate suckering habit.

Leaves – Light green, tip cute, margin entire, surface smooth, 40-48 x 12-15 cm in size.

Rhizome – Creamy-white in colour; 2.0-2.5 cm in dia.; fresh weight varies from 10-15 g per 2.5 cm length, internode length

– 1.5-2.5 cm; no scales; roots 0.3 -0.6 cm in dia., sparsely present.

B. Floral

Inflorescence – Simple, 25-30 cm long; contains 12-16 flowers.

Flower – 8.0 to 10.0 cm in dia., 9.5-10.0 cm in length; bi-coloured, canary yellow (2) along the margin like an irregular band having 2.0-4.0 mm width covering 15 per cent of the individual petal; rest of the petal (85 per cent) is geranium lake – 20 (a shade of red) like flame of fire; margin slightly frilled (Roy, 2007).

Associated Characters and Cultural Practices

The colour combination of the flowers is like flame of fire, a combination of yellow and red, very attractive and distinct. It's a medium height variety, suitable for bedding purpose, free flowering type; peak flowering season is February to October. The cultivar also performs well when grown in pots and suitable for growing in tropical/sub-tropical conditions.

Canna Varietal Evaluation in Sodic Soil

Another set of *Canna* germplasm conservation and evaluation trial was planted at our new established Gehru Centre of Distant Research Centres under sodic soil conditions in 2013. Two experiments, one is on nutrient management, incorporating bio-fertilizers and second on varietal evaluation in sodic soilwas conducted. In one experiment, growth of *Canna* (*var.* Red dazzler) was compared under sole application of chemical fertilizers with combined application of farm yard manure (FYM) along with phosphorus solubilizing bacteria (PSB), *Trichoderma* and combination application of PSB and *Trichoderma*. The response of bio-fertilizers on the growth of *Canna* was found not significant during the investigation. However, response of FYM supplemented with combination of both, PSB and *Trichoderma* were found better than application of chemical fertilizers.

In another experiment, five *Canna* varieties of CSIR-NBRI collection were examined to screen the best suited cultivar for sodic soil. The initial properties of the soil were pH- 9.51 EC- 0.90 dS/m and OC (per cent) - 0.31.The varieties include Caltolen, Pink sunrise, Facceda, Blacknight and Bengal tiger. Significant variation has been seen in terms of their growth performance. Among the varieties significantly higher plant height was recorded in Black night throughout the year. This was followed by Bengal tiger and Cattleya, which were at par with Pink sunrise. However, number of tillers per plant was significantly higher in later stage of growth at 300 DAP (days of planting).

The varietal evaluation was further extended by adding three species having 27 varieties (*Canna maculata* - City of Portland, Red president, Burgundy blush, Angel pink, Cleopatra, New type; *Canna latifolia* - Ambassador, Golden Lucifer, Tropical sun rise, Apricot dream, Assault, Golden Giril, Allegheny, New Delhi yellow, Bangkok, Raktima, Orange punch, King City of Gold, Panama, Rose Mouldcoles, Pink Sunburst, Sinduri dwarf and *Canna lambantum*-Bengal Tiger,Black Night, Cattleya, Flaccida, Pink Sunrise, Red dazzler in September 2013.

Conclusion

The study thus concluded that the response of bio-fertilizers on the growth of *Canna* was found not significant during the investigation. However, response of FYM supplemented with combination of both, PSB and *Trichoderma* were found better than application of chemical fertilizers. Significant variations have been seen in terms of their growth performance like plant height, number of leaves, tillers and flowers per plant. Significantly higher plant height was recorded in Black night throughout the year. This was followed by Bengal Tiger and Cattleya, which were *at par* with Pink Sunrise. However, number of tillers per plant was significantly higher in later stage of growth at 300 DAP (days of planting) under partially sodic waste land.

Acknowledgements

The author is grateful to Director, CSIR- National Botanical Research Institute, Lucknow for encouragement and help and AGTECH project (BSC 0110) for financial support.

For further readings

Dunkley C. S. and Barrett S. 2001. Case Study of the Blue and John Crow Mountains National Park, Jamaica *Canari Technical Report, 282.*

Abrol I.P., Yadav J.S.P. and Massoud F.I. 1988. Salt-affected soils and their management, Soil Bulletin No. 35, FAO, Rome, 131p.

Ashton P.S. 1984. Botanic gardens and experimental grounds. Pp. 39–48 in V. H. Heywood, editor; and D. M. Moore, editor., eds. Current Concepts in Plant Taxonomy. Academic Press, London.

Tisdell C. A. 2011. Core issues in the economics of biodiversity conservation. *Annals of the New York Academy of Sciences*, 1219(1): 99-112.

Kjaer E. D., Graudal L. and Nathan I. 2001. Ex Situ Conservation of Commercial Tropical Trees: Strategies, Options and Constraints, Danida Forest Seed Centre, Humlebaek, Denmark.

Frankel O.H. and Soule M. E. 1981. Conservation and Evolution. Cambridge University Press, New York. 327 pp.

Glowka L., Guilmin F. B. and Synge H. 1994. A Guide to the Convention on Biological Diversity *Environmental Policy and Law Paper No. 30* IUCN Gland and Cambridge. Xii + 161 pp.

https://www.indiaagronet.com/horticulture/CONTENTS/cannas.htm

Antofie M. 2011. Current political commitments' challenges for ex situ conservation of plant genetic resources for food and agriculture," *Analele Universitãþii din Oradea–Fascicula Biologie*, 18: 157-163.

Sharma D.K. and Chaudhari S.K. 2012. Agronomic research in salt-affected soils of India: An Overview. *Indian Journal of Agronomy*, 57:175-185.

Borokini T.I., OkereA. U., Giwa A. O., Daramola B.O. and Odofin W. T. 2010. Biodiversity and conservation of plant genetic resources in Field Gene-bank of the National Centre for Genetic Resources and Biotechnology, Ibadan, Nigeria," *The International Journal of Biodiversity and Conservation*, 2: 37-50.

Chapter 14

Germplasm Conservation of different *Canna* Cultivars at National Botanic Garden, Trombay (NBGT), Mumbai

C.K. Salunkhe and H.A. Barbhuiya

Landscape and Cosmetic Maintenance Section,
Architectural and Structural Engineering, Division,
Bhabha Atomic Research Centre, Trombay, Mumbai – 400 085, Maharashtra
E-mail: chandrak@barc.gov.in, hahmed@barc.gov.in

Abstract

In the present study we have found that NBGT is conserving living germplasm of ca. 60 Canna cultivars for various ornamental purposes and sustainable gardening. To determine their growth and flowering characteristics in Mumbai condition 54 selected cultivars were subjectively evaluated for rate of spread, floriferousness and seed set. Based on their performance in the field trial cultivars were broadly categorized into three groups viz. tall (14), medium (24) and low-growing (16) varieties. Besides above, significant morphological characters of each cultivar were presented in tabular form.

Keywords: Canna, Ornamental plant, Field trial, Propagation, Phenology.

Introduction

Cannas or "Indian Shots" are one of the most versatile ornamental plants valued for its foliage and flowers. It's alluring beautiful flowers may be of colour yellow, orange, pink or red and large banana-like green, bronze-burgundy or variegated foliage attracts eyes of many common peoples as well as of both botanists and horticulturists which instigate to grow them at various gardens and landscapes. The genus *Canna* L. belongs to the monogeneric family Cannaceae Juss. comprising of about 10 species, distributed all over the Neotropics, from Virginia in the USA to northern Argentina and are widely cultivated all over the Tropics and Subtropics, and often escaped and naturalized (Mass-Van de Kamer and Mass, 2008). *Canna* are rhizomatous herbaceous perennials with erect stems bearing ovate leaves, having showy flowers with showy petal-like staminodes and small, coloured petals and sepals, borne in racemes or panicles. Until the mid-nineteenth century, the *Canna* was known chiefly as a starch crop, they were obscure tropical plants- tall, long-jointed and small-flowered. Hybridization of *Canna* for ornamental use was started in 1848 when Thré Année brought some *Canna* species from South America to France. Année's plants were chiefly foliage subjects. In 1870s a French breeder, Antoine Crozy, succeeded in developing larger flowered cultivars. Towards the end of the nineteenth century, Carl Sprenger in Italy and Wilhelm Pfitzer in Germany were continuing to produce novelties and all these varieties fuelled the sub-tropical bedding craze of the past (Gray and Grant, 2003). The hybridized *Canna* has rapidly gained popularity and became the height of horticultural fashion as they have dwarf habit with large, bold inflorescence of large highly coloured flowers (Cooke, 2001).

In India cross breeding of *Canna* was first started in Lucknow and the first hybrids were recorded by Mr. Percy Lancaster in 1890 and by Mr. Ridley, Superintendent of the Govt. Horticultural Garden (now National Botanical Research Institute, Lucknow) in 1892. Others who worked on the *Canna* were Mr. A.J. Heins of the Lahore Government Gardens and Mr. H.E. Houghton of the Agri-Horticultural Society's Garden, Madras. Work was also carried out in Calcutta from 1892 till 1953 by the Lancasters, father and son (Percy-Lancaster, 1957).

Most of the present day cultivars are complex hybrids. *Canna indica* and *C. glauca* are the principal progenitors of the multiple lineages of cultivated *Canna's*. Breeders have also used *C. flaccida*, *C. iridiflora*, and *C. liliiflora* many times to create new forms with larger flowers. Bailey (1923) made an attempt to provide a binomial for cultivated *Canna's*, describing *Canna generalis* and *C. orchiodes*, as *'cultigen novum'*. He characterized former by erect petals and a short floral tube, whereas later by reflexed petals and a long floral tube. However, these names were not accepted as a valid hybrid name by Mass-Van de Kamer and Mass (2008), as because parent species names were not mentioned. Moreover, they included both the species under synonymy of *C. indica*, because of the fact that Bailey and Bailey (1930) added to their description "Of garden origin" does not necessarily mean these names do only refer to garden plants.

Botanic gardens have a long tradition of introducing new plant cultivars into cultivation and maintaining them for long periods and in this aspect the role of NBGT has no exception. Since its inception many new cultivars were introduced for various types of gardening and landscaping. To preserve living germplasm of various *Canna* cultivars NBGT has established a *Canna* garden in its campus, which spread over 15000 sq. m. area (Figure 14.1). So far, *ca*. 60 *Canna* cultivars are conserved in this garden for various requirements in landscaping and sustainable gardening.

Methodology

About 10–15 cm long rhizome pieces (each with a growing point) of 54 *Cannas* cultivars were planted in field spreading over 550 sq. m area. Each rhizome piece was spaced 1 m apart in rows and each row was spaced 1.5 m apart and rhizomes buried at a depth of 7–8 cm under the soil. Six replicates of each cultivar were planted in a completely randomized experimental design. All plants were fertilized in every 3 months with fertilizers containing all essential nutrients and were kept weed free by mulching and by spraying weedicide. After that, when plants reached an equilibrium height, the heights were measured and evaluated subjectively for rate of spread, rate of flowering and seed set on three dates during 2015.

Results and Discussion

Canna cultivars evaluated are divided into 3 major categories based on average total height of the plants growing under Mumbai condition: tall varieties (>130 cm), medium varieties (90-130 cm) and low growing varieties (<90 cm). All the three categories are discussed below (Figure 14.2).

Tall Varieties

In this category 4 bronze leaves, 10 green-leaves cultivars were tested (Table 14.1). Among bronze leaved cultivars, 'Black Knight' grew the tallest, spread at a fairly rapid rate, produced red flowers consistently, and produced no seeds. The cultivar 'Soundarya' has very robust in habit as well as have attractive dark-bronze foliage, produces red flowers and attains moderately to fairly tall height. Its rate

Figure 14.2: Some Beautiful Canna Cultivars Conserved at National Botanic Garden, Trombay.

of spread, flowerings are also high and have no seeds set. The cultivar 'Umbria' is tallest among green-leaved cultivars, which spread at a medium rate. It produces red flowers (with yellow margins on staminodes) at a medium rate and have no seed set.

Table 14.1: Growth and Flowering Characteristics of Tall (>130 cm) *Canna* Varieties at NBGT, Mumbai (z0 = No spread, z3 = Very aggressive; y0 = No flowers, y3 = Very floriferous; x0= No seed set, x3 = Seeds always set)

Sl. No.	*Name of the Cultivar*	*Foliage*	*Flower Colour*	*Height in cm*	*Rate of Spread*[z]	*Rate of Flowering*[y]	*Rate of Seed Set*[x]
1	*C.* 'Black Knight'	Bronze	Red	140-207	3	3	0
2	*C.* 'Soundarya'	Bronze	Red	131-184	3	3	0
3	*C.* 'Statue of Liberty'	Bronze	Red	132-206	3	3	0
4	*C.* 'Wyoming'	Bronze	Orange	130-185	3	2	0
5	*C.* 'American Beauty'	Green	Red	140-208	3	3	0
6	*C.* 'Autumn Tint'	Green	Yellow with orange patches and spots	130-184	3	2	0
7	*C.* 'Countess of Wellingdon'	Green	Red	167-209	3	2	0
8	*C.* 'Hungaria'	Green	Red	130-194	3	3	0
9	*C.* 'Mrs. Pierre S. DuPont'	Green	Pink	140-185	2	2	0
10	*C.* 'New Red'	Green	Red	100-240	3	2	0
11	*C.* 'Roi Soleil'	Green	Dark red	130-168	2	2	1
12	*C.* 'Sadia'	Green	Yellow	134-200	3	3	0
13	*C.* 'Umbria'	Green	Red with Yellow margins on staminodes	140-267	2	2	0
14	*C.* 'Unique'	Green	Pink	135-157	2	2	0

Medium Height Varieties

Twenty-four cultivars of medium height (90-130 cm) were evaluated (Table 14.2), of which 4-bronze, 17-green, 1-either green or bronze and are 2-variegated leaved. Among green-leaved varieties 'Apricot', 'Buttercup', 'Colossal' 'Lalarookh' and 'Rosamond Coles' are best varieties for flowering as their growth and height is moderate and also have congested floral axis, hence inflorescence always looks full of flowers.

The cultivar 'Percy Lancaster' is the only exceptional variety. It produces green foliage normally but sometimes it also produces dark-bronze foliage; flowers are usually yellowish and heavily spotted with red. Its rate of spread is high, rate of flowering is moderate and have no seed set.

Table 14.2: Growth and Flowering Characteristics of Medium Height (90-130 cm) *Canna*

Varieties at NBGT, Mumbai (z0 = No spread, z3 = Very aggressive; y0 = No flowers, y3 = Very floriferous; x0= No seed set, x3 = Seeds always set)

Sl. No.	*Name of the Cultivar*	*Foliage*	*Flower Colour*	*Height in cm*	*Rate of Spread*[z]	*Rate of Flowering*[y]	*Rate of Seed Set*[x]
1	*C.* 'Councellor Benoist'	Light Bronze	Light red	98-125	1	2	0
2	*C.* 'Empress of India'	Bronze	Red	76-124	2	3	0
3	*C.* 'Raktima'	Bronze	Red	103-130	1	1	0
4	*C.* 'Red King Humbert'	Bronze	Red yellowish at base	77-175	3	2	0
5	*C.* 'Apricot'	Green	Pink	90-116	3	3	0
6	*C.* 'Buttercup'	Green	Cream/Light yellow	90-130	3	2	0
7	*C.* 'City of Portland'	Green	Red	91-109	2	2	0
8	*C.* 'Cleoparta'	Green	Pink	90-127	3	1	0
9	*C.* 'Colossal'	Green	Red, bordered with yellow	92-117	2	3	0
10	*C.* 'Eastern Queen'	Green	Yellow with orange spots at centre	102-130	1	2	0
11	*C.* 'Edith'	Green	Orange	111-117	2	2	0
12	*C.* 'Fire Bird'	Green	Red	110-129	2	2	1
13	*C.* 'Ideal'	Green	Yellow with red throat	81-118	2	2	1
14	*C.* 'Khusuru Baugh Red'	Green	Red with yellow tinge	75-130	3	2	0
15	*C.* 'Lalarookh'	Green	Yellow with light orange tinged inside	64-131	3	3	1
16	*C.* 'Lucifer'	Green	Yellow with dark red patches	70-135	3	3	1
17	*C.* 'Maharaja of Gwalior'	Green	Pink	51-101	1	1	0
18	*C.* 'Master Piece'	Green	Red	90-100	2	2	0
19	*C.* 'Plume'	Green	Pink	101-123	2	2	0
20	*C.* 'Rosamond Coles'	Green	Yellow with red spots	100-104	2	3	0
21	*C.* 'Yellow Queen'	Green	Yellow	90-127	1	1	0
22	*C.* 'Percy Lancaster'	Green/ Bronze	Yellow, heavily spotted with red	84-133	3	2	0

Sl. No.	Name of the Cultivar	Foliage	Flower Colour	Height in cm	Rate of Spread[z]	Rate of Flowering[y]	Rate of Seed Set[x]
23	*C.* 'Bengal Tiger'	Variegated (green/ yellow)	Orange-yellow	82-130	3	3	0
24	*C.* 'Sun Burst'	Variegated (Bronz/ Green/ Yellow)	Orange	76-130	2	2	0

Low-growing Varieties

Sixteen low-growing cultivars were evaluated (Table 14.3), out of which 2-bronze, 13-green and 1-variegated leaved. 'Pink Dwarf' is the shortest cultivar among all green-leaved variety; this plant always looks bushy and produces pink colour flowers. The rate of spread and rate of flowering is moderate.

The cultivar 'Trinacaria Variegata' is the shortest among variegated-leaved verities, having moderate rate of spread and flowering.

Table 14.3: Growth and Flowering Characteristics of Low Growing (<90 cm) *Canna* Varieties at NBGT, Mumbai (z0 = No spread, z3 = Very aggressive; y0 = No flowers, y3 = Very floriferous; x0= No seed set, x3 = Seeds always set)

Sl. No.	Name of the Cultivar	Foliage	Flower Colour	Height in cm	Rate of Spread[z]	Rate of Flowering[y]	Rate of Seed Set[x]
1	*C.* 'Laglore'	Bronze	Red	40-60	2	2	0
2	*C.* 'Susquehana'	Light Bronze	Pink	48-79	3	3	0
3	*C.* 'Agnishika'	Green	Red	40-78	2	2	0
4	*C.* 'Angels Robe'	Green	Dark pink	65-67	1	2	0
5	*C.* 'Beautiful'	Green	Yellow	54-72	1	1	0
6	*C.* 'Copper Giant'	Green	Yellow with orange tinge	41-90	1	1	0
7	*C.* 'Dainty'	Green	Yellow	71-90	2	1	1
8	*C.* 'Konark'	Green	Cream	70-90	1	2	0
9	*C.* 'Lady Lawrence'	Green	Orange	38-61	2	3	0
10	*C.* 'Moon Light'	Green	Yellow	23-43	1	1	0
11	*C.* 'Orange Queen'	Green	Orange	73-90	3	2	0
12	*C.* 'Pink Dwarf'	Green	Pink	22-38	2	2	0
13	*C.* 'Prime Rose'	Green	Yellow	75-80	2	2	0
14	*C.* 'Rajbhawan'	Green	Pink	40-85	1	1	0
15	*C.* 'Richard Wallace'	Green	Orange	42-62	1	1	0
16	*C.* 'Trinacaria Variegata'	Variegated (green/ yellow)	Yellow	50-88	2	2	0

Conclusion

Within each height category there are usually bronze, green and occasionally variegated-leaved cultivars with red, pink, orange or yellow flowers which are suitable for use as an ornamental plant in Mumbai condition. Generally, rate of spread and floriferousness are correlated with height, the larger varieties being more vigorous overall. Rapidly spreading varieties should be avoided as they are difficult to confine their bed and may form large, sparse, and unattractive clumps, rather than desirable compact clumps. Most of the aggressive varieties also have rather unattractive flowers. Out of the 55 cultivars evaluated following six cultivars *viz.* 'Dainty', 'Fire Bird', 'Ideal', 'Lalarookh', 'Lucifer' and 'Roi Soleil' occasionally produces seeds. Germination of *Canna* seeds in the beds is seldom a serious problem in Mumbai, but the energy diverted for the production of seeds is used in non-seeding plants to produce more flowers. The trend in recent years has been to plant dwarf cultivars with more stems per rhizome. These plants tend to be bushy and very floriferous, but those who desire the traditional, taller, more imposing varieties still have plenty to choose from, and even the tallest rarely require support.

Acknowledgements

Authors are thankful to Shri K.N. Vyas, Director, BARC, for providing facilities. We also extend our sincere thanks to Shri R. Narasimhan, Head, A and SED and Shri H. Mishra AD, ESG, for their support.

References

Bailey L.H. 1923. Various cultigens, and transfers in nomenclature. *Gentes Herbarum* 1: 118–120.

Bailey L.H. and Bailey E.Z. 1930. *Hortus, a concise dictionary of gardening, general horticulture and cultivated plants in North America*. Macmillan Company, New York.

Cooke I. 2001. *The Gardener's Guide to Growing Cannas*. Timber Press, Portland.

Gray J. and Grant M. 2003. *Canna: RHS Plant Trials and Awards*. Royal Horticultural Society, Wisley.

Mass-Van de Kamer H. and Mass P.J.M. 2008. The Cannaceae of the world. *Blumea* 53: 247–318.

Percy-Lancaster S. 1957. Canna. *Bulletin of National Botanic Gardens, Lucknow* 3: 1–15.

Chapter 15

Bougainvillea at CSIR-NBRI Botanic Garden: Germplasm Collections, Agro-technology and Development of New Varieties

Shilpi Singh and R.K. Roy

Botanic Garden, Floriculture and Landscaping Division,
CSIR-National Botanical Research Institute, Lucknow, Uttar Pradesh
E-mail: ss_lifescience@yahoo.com

Abstract

Bougainvillea belongs to the family Nyctanginaceae and is one of the most popular ornamental plants in tropical and sub-tropical gardens. This crop is much popular for an ornamental purpose all over the world for their colourful bracts available in a wide range of colours. CSIR-NBRI was the first institute to take up research work on Bougainvilleas. Research work on Bougainvillea in NBRI was first initiated by S. Percy Lancaster. CSIR-NBRI, Lucknow is having very rich germplasm of Bougainvillea. At present, the institute has collection of four basal species of Bougainvillea (B. spectabilis, B. glabra, B. peruviana and B. x buttiana) which have horticultural importance and 200 varieties. This institute developed 26 number of new and novel varieties by employing different plant breeding methods. The living germplasm collection of Bougainvilleas has been maintained in the two ways (potted and ground). Bougainvilleas have prominent landscaping uses also.

Keywords: *Bougainvillea, Bracts, Germplasm, Varieties, Novel*

Introduction

Bougainvillea is a very popular ornamental plant grown all over the world for their colourful bracts available in a wide range of colours *viz.* white, yellow, pink, red, mauve, bi-coloured and multi-coloured. Moreover, leaves of some of the varieties have unique variegation in various colours and combinations. The colourful and attractive foliage is also an added attraction particularly when the plant is out of bloom.

An easy adaptability to different agro-climatic conditions and free flowering habit, have made Bougainvilleas obvious choice for every garden. The plant withstands well in adverse climatic and stress conditions due to hardy nature and produce colourful bracts as well. The requirement of maintenance is also less in comparison other garden plants. A large number of varieties is available in the horticultural nursery trade with varied growth habit and ideal for growing in the gardens for various purposes. Considering overall performances, *Bougainvillea* are most suitable flowering plant for the gardens of sub-tropical and tropical regions of the world.

Bougainvillea belongs to the family Nyctanginaceae and is one of the most popular ornamental plants in tropical and sub-tropical gardens. Due to high popularity in the tropical countries, *Bougainvillea* is called 'Glory of the Tropics'. From its place of origin - Rio De Jenerio, Brazil, the plant has travelled a lot and domesticated in almost all parts of the world. However, growth, proliferation and development of bract colour are best suited to tropical and sub-tropical climate. Therefore, the popularity of Bougainvilleas is much more in the tropical parts of the world compared to others. The French Botanist, Commerson first collected the plant from Rio de Janerio, Brazil (Golby, 1970).

B. spectabilis was the first species introduced in India from Royal Botanic Garden, Kew in 1860. *B. glabra* was introduced to Agricultural and Horticultural Society of India at Alipore, Calcutta in 1869 (Anonymous, 1894). This society played a significant role in the establishment of horticulture trade in ornamental plants. (Lancaster, 1959; Sharma, 1996). Bougainvilleas are the most floriferous shrubs for producing beautiful colour effects which draw attention of the garden lovers.

Germplasm Collection

CSIR-NBRI, Lucknow is having very rich germplasm of *Bougainvillea*. The living germplasm collection of Bougainvilleas has been maintained in the following two ways.

Potted Collection: This collection has been displayed and is being maintained in an enclosure measuring 50x8 m. The varieties have been arranged and displayed in alphabetical order keeping three potted plants together. Altogether, there are 600 pots in the collection having plant labels so that one can easily identify and locate the varieties.

Figure 15.1: Bougainvillea Germplasm Collection at CSIR-NBRI.

Field Collection: This collection has been developed as a *Bougainvillea* Garden over an area of one acre. There are 165 varieties planted in this garden as per specific layout plan. There are four sectors planted with four horticulturally important species *viz. B. glabra, B. spectabilis, B. peruviana* and *B. x buttiana.* Under these species different varieties have been planted in rows at a distance of 5 m. All the varieties are labelled for their educative and aesthetic use.

Figure 15.2: Bougainvillea Garden in CSIR-NBRI.

Methods for Development of New Varieties

Bud Sport/Spontaneous Mutation

These are created naturally without any human interference. It is required to spot these natural mutants out of the whole germplasm collections. There may be changes of flower colour, foliage aberration or any other change in morphological characters. Bud sports are isolated, multiplied vegetatively and performance is evaluated in the subsequent generations with regard to the stability of changed morphological characters and its distinctiveness in comparison to the parental characters (Zadoo, *et al.*, 1975a).

Chemical Mutagenesis

Application of chemical mutagens like Ethyl Methane Sulphonate EMS (0.01 - 0.03 per cent) for induction of somatic mutations (foliage/flower). These are mainly used as aqueous solution in various concentrations and applied over the current vegetative growth (meristematic cells) for creating genetic variability either in vegetative part or in the flowers. The mutants thus developed are separated out subsequently, multiplied and ultimately developed into new variety after evaluation through standard protocol (Jayanti and Datta 2006).

Hybridization

Crossing between selected parents having desirable characters. The hybridization technique involving emasculation and pollination specifically for *Bougainvillea* was elaborated by Percy-Lancaster (1959), Zadoo and Khoshoo (1975). Morphological expression may have various changes with respect to flower colour, size, flowering season, *etc.*

Mutagenesis

Induced mutagenesis by application of Gamma Ray (Cobalt 60) for genetic manipulation leading to new cultivars. Usually stem cuttings in required length are irradiated with different doses and planted for root regeneration and growth (Gupta, 1972; Datta 2009).

R&D work at CSIR-NBRI on *Bougainvillea*

Multidisplinary research work was initiated at NBRI for the improvement of Bougainvilleas through hybridization, selection, polyploidy and mutation breeding. As a result, a large number of new and novel varieties developed by the Institute. In India, CSIR-NBRI was the first institute to take up research work on Bougainvilleas. Research work on *Bougainvillea* in NBRI was first initiated by S. Percy Lancaster. He developed very novable varieties. Later R&D work was done by T.N. Khooshoo and his associates and as a result very promising varieties have been developed. Collection of varieties was initiated from various sources and conserved as living specimens. At present, the institute has collection of four basal species of *Bougainvillea* (*B. spectabilis, B. glabra, B. peruviana* and *B. x buttiana*) and 200 named varieties. These collections have been recognized as 'National Collection' by Ministry of Environment, Forest and Wildlife; Ministry of Agriculture, Govt. of India and National Biodiversity Board. Moreover, these collections are often referred for identification, research and improvement work on *Bougainvillea* taken up by National and International institutions/organizations. (Roy, 2015; Banerji, 2012; Sharma and Roy, 2002; Banerji and Datta, 1986).

Development of New Varieties

The broad genetic base developed in the form of live germplasm collection was used for various research work for the development of new varieties through hybridization, selection, bud sport, polyploidy and mutation breeding. As a result, a large number of new and novel varieties have been developed by the institute. Altogether, 26 new varieties have been developed by employing different plant breeding methods (Table 15.1).

Table 15.1: List of Varieties Developed by CSIR-NBRI

Sl. No.	*Name of the Varieties*	*Methods*	*Parent*	*Year of Development*
1	'Begum Sikander'	Hybridization	'Dr. B.P. Pal' x 'Jennifer Fernie'	1969
2	'Shubhra'	Bud Sport or Spontaneous Mutation	'Mary Palmer'	1969
3	'Dr. B.P. Pal'	Colchiploidy	'Shubhra'	1969
4	'Tetra Mrs. McClean'	Colchiploidy	'Mrs. McClean'	1969
5	'Arjuna'	Induced Mutation (Gamma Radiation)	'Partha'	1972
6	'Archana'	Bud Sport or Spontaneous Mutation	'Roseville's Delight'	1973
7	'Mary Palmer Special'	Hybridization	'Princess Margaret Rose' x 'Dr. B.P. Pal'	1974
8	'Wajid Ali Shah'	Hybridization	'Dr. B.P. Pal' x 'Mrs. Chico'	1974
9	'Parthasarthy'	Bud Sport or Spontaneous Mutation	'Partha'	1977
10	'Shweta'	Bud Sport or Spontaneous Mutation	'Trinidad'	1979
11	'Chitra'	Hybridization	'Tetra Mrs. McClean' x 'Dr. B.P. Pal'	1981
12	'Surekha'	Bud Sport or Spontaneous Mutation	'Scarlet Queen'	1981
13	'Nirmal'	Bud Sport or Spontaneous Mutation	'Mrs. McClean'	1982
14	'Manohar Chandra Variegata'	Bud Sport or Spontaneous Mutation	'Manohar Chandra'	1985
15	'Pallavi'	Induced Mutation (Gamma Radiation)	'Roseville's Delight'	1987
16	'Hawaiian Beauty'	Bud Sport or Spontaneous Mutation	'Hawaiin White'	1990
17	'Los Banos Variegata'	Induced Mutation (Gamma Radiation)	'Los Banos Beauty'	1990
18	'Mahara Variegata'	Induced Mutation (Gamma Radiation)	'Mahara'	1994
19	'Los Banos Variegata - Silver Margin'	Induced Mutation (Gamma Radiation)	'Los Banos Beauty'	2002

Sl. No.	Name of the Varieties	Methods	Parent	Year of Development
20	'Mahara Variegata Abnormal Leaves'	Induced Mutation (Gamma Radiation)	'Mahara'	2002
21	'Los Banos Variegata Jayanti'	Chemical Mutagen Induced Mutants	'Los Banos Beauty'	2006
22	'Aruna'	Induced Mutation (Gamma Radiation)	'Palekar'	2008
23	'Pixie Variegata'	Chemical Mutagen Induced Mutants	'Pixie'	2009
24	'Abhimanyu'	Bud Sport or Spontaneous Mutation	'Arjuna'	2010
25	'Dr. P.V.Sane'	Induced Mutation (Gamma Radiation)	'Dr. R.R.Pal'	2011
26	'NBRI-Dr. A.P.J. Abdul Kalam'	Bud sports	Fantasy	2015

Agrotechnology

Climate and Soil

Bougainvilleas perform best under full sun and in hot climate. They can withstand wide range of climatic conditions, but growth and development of bracts are affected under adverse conditions. In north Indian conditions, best flowering period is during April –June. During this period, the average maximum day temperature ranges between 35-40°C and the relative humidity 35-55 per cent.

Bougainvilleas can be grown in all types of soil. However, the soil should be well drained and rich in plant nutrients. They do not grow well in wet soil as the water requirement is comparatively less. Slightly acidic soil having pH 6.0-6.5 is good for growth and flowering as it helps in availability of micro-nutrients.

Planting

Well grown and healthy plants are generally selected for planting. In eastern and northern India, planting during February to April and subsequently in July to September. In southern India, planting can be done almost round the year particularly when the temperature is not very high and level of humidity is adequate.

In Ground: Planting in ground is done along the fence/boundary wall, as specimen plant, training over the gate, porch, pergola and several other ways. Sometimes, planting in group is also done for achieving mass effect of bract colour. Generally 60x60x60 cm pits are prepared for the ground plantation. Planting distance depends on the purpose and usually kept 2-3 m in between plants.

In Pots: Bougainvilleas grow well in pots or in large containers and produce flowers well. The main advantage is that the pots/containers can be placed for decoration purpose in any part of the gardens. The size of the earthen pots/cement

Figure 15.3: Dr. P.V. Sane (CSIR-NBRI New Variety-2011).

Figure 15.4: Dr. APJ Abdul Kalam (CSIR-NBRI New Variety-2015).

pots should be 25 to 90 cm in diameter used for growing specimen plants. Planting in pots/containers is done in moderate season of the year either in February -March or July-September.

Propagation

Propagation of Bougainvilleas is done in various ways. Stem cutting and layering are usually followed for large scale propagation. The different propagation methods usually practiced for Bougainvilleas are described below.

Stem Cuttings: In this method stem cuttings with hardwood having diameter of 0.75 to 1.5 cm are selected. Tip of the cuttings is given slanting cut while the basal portion is given round cut just below the node. Rooting of the cuttings is done in various ways either putting them in well prepared soil/sand beds or in containers. Stem cutting are treated with auxin (hormone) powder like IAA (Indole Acetic Acid) (Hackett, *et al.,* 1972).

Layering: This is another type of vegetative propagation suitable for those varieties which does not respond well to stem cuttings. Stem having pencil-thickness are selected for air-layering. A simple slanting cut is given followed by wrapping with sphagnum moss, covered with polythene sheet and tied by cord. When adventitious roots are seen through polythene sheet, the stem is cut from the mother plant and planted in pots/containers as new plant.

Budding: This is an alternative method of propagation but not done in large scale. Varieties which do not respond well by stem cutting are usually propagated by budding. In north-India the best period for budding is March-April but can be done in any season except monsoon.

Pruning

Pruning refers removal of vegetative part (main branch, sub-branch and then shoots) of the mother plant. The requirement of pruning and training depends on the use of the plant *i.e.,* hedge, cascade, espalier, pot plant *etc.* If it is grown as climber on porch, arch, on stump, pruning and training becomes essential. Over grown, dead and de-shaped branches are required to be pruned preferably before monsoon or after the flowering.

Manures and Fertilizers

Established plants should be manured twice in a year once in July and in February for encouraging growth and flowering. Bougainvilleas have modest requirement of manures and fertilizers in comparison to other flowering perennial plants. Grown up and well established plants hardly require application of manures and fertilizers (Iredell, 1990).

For ground plants annual application of mixture of FYM (5-7 kg), bone meal (250 g), neem cake (250g), superphosphate (250g) and potash (50 g) per plant is recommended. The mixture should be applied in the ground around the basal ring followed by irrigation. Top dressing of the above mixtures may be done at regular interval before forking of the plants.

In pot/container grown plants, application and forking of the manure mixture is recommended especially after pruning and in active growth period. The mixture for pots should be prepared properly by mixing garden soil and FYM in 3:1 ratio together with Neem cake and bone meal @100 g/pot respectively (Singh and Dadlani, 1986).

Irrigation

Young plants in ground require watering at the initial stage for better establishment. Established plants in ground, however, require less watering. Irrigation during pre-flowering and flowering stage is to be restricted which encourage profuse flowering. In pots young plants require regular watering till they are established. Frequency of irrigation is decided according to season. Overwatering should be avoided during flowering season as it may result in to poor or lack of flowering.

Diseases and Pests

Bougainvilleas are very hardy and are not attacked by diseases and pest in large scale. All over the world, there is no report regarding major diseases and pest which have been found to limit growth and flowering of Bougainvilleas. However, there are some report of minor diseases and pests which have been found affecting the plant occasionally. Very rarely viral, bacterial and fungal infections are seen in this crop. No major pests have been reported which attack Bougainvilleas in alarming scale. Sometimes, caterpillars, aphids, leaf miners, scale insects, thrips, white fly *etc.* have been found to attack the plant. These can be controlled by spraying of systemic insecticides as and when required.

Landscape Use

Bougainvilleas are used in the landscaping in a large scale and in various ways. Their varied adaptability and growth habit facilitate usages in different styles. The most popular form of growing is as pot plant besides ground plantation for various purposes. Some of the most prominent uses of Bougainvilleas in landscaping have been used *viz.* Arch or Pergola, Bonsai, Bush, Climber, Espalier, Hanging basket, Hedge, Cascade, Slopes and Mound, Standard, Training on tree or stump, Planting along the road, Central Verge, Traffic Island, Window Sill, *etc.* (Roy, 2008).

References

Anonymous. 1894. *Bougainvillea glabra* var. 'Sanderiana'. *Proceeding of Royal Horticulture Society.* 17:8.

Banerji, B.K. 2012. Improvement of Bougainvilleas. Approaches and Methods. *Indian Bougainvillea Annual* 24:17-22

Banerji, B.K. and Datta S.K. 1986. Improvement of ornamental plants by induced mutation: *Bougainvillea*. *Floriculture Today* 3:4-9

Datta, S.K. 2009. A Report on 36 Years of Practical Work on Crop Improvement Through Induced Mutagenesis: Induced Plant Mutations in the Genomics Era. Food and Agriculture Organization of the United Nations, Rome 253-256.

Golby, Eric V. 1970. History of *Bougainvillea* in Florida (PP. 138-45), *In* Edwin A. M. Flowering vines of the world: An Encyclopedia of Climbing Plants. Hearthside Press Inc., Publishers, New York.

Gupta, M.N. 1972. Mutation breeding of some vegetatively propagated ornamentals. In: Progress in Plant Research, NBRI Silver Jubilee Publication, Today and Tomorrow's Printers and Publishers, New Delhi, pp 75.

Hackett, W.P., Sachs, R.M. and DeBie, J. 1972. Growing *Bougainvillea* as a flowering pot plant. *California Agriculture* 26 (8): 12-13.

Iredell, J. 1990: The *Bougainvillea* Growers Handbook. Simon and Schuster, Brookvale, New South Wales, Australia, pp 111.

Jayanti, R. and Datta, S.K. 2006. Improvement of *Bougainvillea* varieties using Chemical Mutagens. In: *National Conference on Bougainvillea*, April 12-13, 2006, NBRI, Lucknow, pp17.

Lancaster, S.P. 1959. *Bougainvillea*. Bull. 41, National Botanical Gardens, Lucknow, 31 pp.

Roy, R.K., Singh, S., and Rastogi, R.R. 2015. *Bougainvillea*: Identification, Gardening and Landscape Use (Lucknow: CSIR-National Botanical Research Institute), pp.144.

Roy, R.K. 2008. *Bougainvillea*–An excellent ornamental for every garden. *Indian Bougainvillea Annual* 21: 18-20.

Sharma S.C. and Roy, R.K. 2002. Conservation and improvement of *Bougainvillea* at NBRI. *Floriculture Today* VI (8): 8-9.

Sharma, S.C. 1996. Bougainvilleas in India (Lucknow: EBIS, National Botanical Research Institute), pp.74.

Singh, B. and Dadlani, N.K. 1986. *Bougainvillea*. In: Ornamental Horticulture: India (Eds. Chaddha and Chaudhury). IARI, New Delhi, 10-12.

Zadoo, S.N. and Khoshoo, T.N. 1975. Nature of self-incompatibility in cultivated Bougainvilleas. *Incompatibility Newsletter*, 5: 73-75.

Zadoo, S.N., Roy, R.P. and Khoshoo, T.N. 1975a. Cytogenetics of cultivated *Bougainvillea* - III: Bud sports. *Z. Pjlanzenzuchtg*, 74: 223-39.

Chapter 16

Fragrant Tuberoses for Enhancing Beauty of Botanic Gardens

P. Ranchana and M. Kannan

Department of Floriculture and Landscaping, Horticultural College and Research Institute, Tamil Nadu Agricultural University, Coimbatore – 641 003
E-mail: ranchanahorti@gmail.com

Abstract

Tuberose (Polianthes tuberosa Linn.) *is one of the important bulbous flower crops of tropical and subtropical regions which produce waxy white flowering spikes of single as well as double types of flower that impregnate the atmosphere with their sweet fragrance and possesses long vase life. These flowers have long been used in perfumery as a source of essential oils and aroma compounds. It's essential oil is exported at an attractive price to France, Italy and other countries as long as there is no synthetic flavour to replace its fragrance. It is also widely grown as specimen plant for exhibition and also as a cut flower. Flowers remain fresh for a pretty long time. The botanic garden of Tamil Nadu Agricultural University, Coimbatore maintains a germplasm collection of tuberose which includes cultivars from various regions like Calcutta (Calcutta Single and Calcutta Double), Hyderabad (Hyderabad Single and Hyderabad Double), Assam (Kahikuchi Single), Gujarat (Navsari Local), Pune Single (Pune), Mexico (Mexican Single and Pearl Double) and also hybrids which are released from various institutes of India viz., MPKV, Rahuri, Maharastra (Phule Rajani), IIHR, Bangalore (Prajwal, Shringar, Suvasini, Vaibhav and Arka Sugandhi) and NBRI, Lucknow (Rajat Rekha and Swarna Rekha). Various collections of tuberose are useful in Botanic Gardens for planting in beds and borders. The fragrance emitted during flowering season use to attract visitors. Single types produce more fragrance than double ones. In general, aesthetic value of garden can be improved by growing tuberose flowers. These collections not only helps for exhibiting display in the garden, but also useful for taking up research work in crop production, crop improvement and post harvest studies. Tuberose growing in botanic garden thus played a vital role, directly in creating aesthetic look and indirectly in educational activities.*

Keywords: *Tuberose, Single types, Double types, Cut flowers, Aesthetic, Education.*

Introduction

Tuberose (*Polianthes tuberosa*) occupies an important position among commercial ornamental bulbous crops, because of its highly fragrant flowers which can be used in various ways and is essentially a florists flower. The white tubular flowers emit the most popular odour and hence are cultivated on a large scale in some parts of the world for the extraction of its highly valued natural flower oil. The lingering delightful fragrance and excellent keeping quality are the predominant characteristics of this crop. It is popularly known as 'Rajanigandha', derives its generic name from the greek word *Polios*, which means white or shinning and *anthos* meaning a semi perennial bulbous plant. It is commercially cultivated for cut flowers and loose flowers trade and also for extraction of its highly valued natural flower oil. The serene beauty of the flower spikes, bright white flowers, sweetness of blooms and delicacy of fragrance of this ornamental crop, transform the entire area into a nectarine and joyous one. Fresh flowers and value added products of tuberose are exported from India to USA, Germany, United Kingdom, Italy, The Netherland, Japan, United Arab Emirates and Saudi Arabia.

Importance and Uses

Tuberose is grown for garden decoration in pots, beds and borders. The flowers of the plant are durable although brittle and remain fresh for a pretty long time and stand lond distance transportation due to their waxy nature. The long flower spikes are excellent cut flowers when arranged in vases, bouquets and bowls. The individual florets or the loose flowers are used for making garlands, to make ornaments for bridal make- up and for button- holes. The flowers are used in Java in vegetable soups. Tuberose flower oil is one of the most valuable and expensive perfumers raw material. The yield of concrete from fresh flowers is 0.08- 0.11 per cent of which 18- 23 per cent constitutes alcohol soluble absolute. In exceptional cases, tuberose oil is used in flavouring candy, beverages and baked foods. The oil has been reported to be added in the proportion in the times like non- alcoholic beverages 0.26 ppm; ice creams, ice, *etc.* 0.46 ppm; candy 1.5 ppm and baked foods 1.7 ppm.

The leaves, flowers, bulbs and roots contain sterol, carbohydrates, saponins and traces of alkaloids. Flavonoides are present in leaves and flowers. The most common constituents of tuberose concrete are geraniol, nerol, benzyl alcohol, methyl benzoate, methyl silicate,ethanol, benzyl benzozte, methyl anthranilate and butyric acid. The bulbs contain an alkaloid lycorin which causes vomiting. Two steroidal sapogenins like necogenin and tigogenin have also been isolated from bulbs. Presence of a transfractosidase has also been observed. The flowers emit a delightful fragrance and are a source of tuberose oil. The renowned perfume like 'Poison' is manufactured by using tuberose oil. The fragrant flowers are added along with stimulants or sedatives to popular beverages prepared for chocolate and served either cold or hot. The essential oil is used in non- alcoholic beverages, ice- creams, candy and baked produces. The bulbs are considered to be hot, dry, diuretic and emetic. They are rubbed with turmeric and butter and applied as paste to remove red pimples of infants. The bulbs after being dried and powdered, are used as remedy for gonorrhoea.

Origin

Tuberose is native to Mexico from where it spread to different parts of the world during 16th century. The name tuberose is derived from tuberosa, the plant being the tuberous hyacinth as distinguished from bulbous hyacinth. The name, therefore, is Tuberose. By the middle of the 18th century, the tuberose was still regarded to as *Hyacinthus indicus tuberosus*. In 4th edition of the Gardeners dictionary printed in 1743 and written by Philip Miller, two species are cited: *Hyacinthus indicus tuberosus flore*, the Indian Tuberose and *Hyacinthus indicus tuberosus flore pleno.* Originally Linnaeus gave the name *Polianthus floribus alternis* to this plant

Distribution

Tuberose is cultivated on large scale in France, Italy, South Africa, North Carolina in USA and in many tropical and subtropical areas including India. In India tuberose is commercially cultivated over 3000 hectare area. The major tuberose growing states in India are presented in Table 16.1.

Table 16.1: Tuberose Growing Major States in India

State	*District*
Assam	Guwahati, Jorhat, Hatikhuli, Tinsukia and Dibrugarh
Maharastra	Pune, Nasik, Satara and Ahmednagar
Gujarat	Navsari, Valsad, Surat and Baroda
Haryana	Ambala, Gurgoan, Faridabad and Hissar
Karnataka	Tumkur, Kolar, Belgaum, Mysore and Bangluru Rural
Andhra Pradesh	Guntur, Chitoor, Krishna, Khadappa and Ranga Reddy
Tamil Nadu	Coimbatore, Madurai, Theni, Trichy and Dindigul
Uttar Pradesh	Meerut, Ghaziabad, Muzaffarnagar, Saharanpur, Lucknow, Kanpur, Kannauj and Barabanki
Uttarkhand	Udham Singh Nagar, Haridwar and Dehradun
Orissa	Cuttak, Puri and Benjam

Morphology

Tuberose is a half- hardy, bulbous perennial perpetuating itself through bulblets. Bulbs are made up of scales and leaf bases and the stem is a condensed structure which remains concealed within scales. Roots are adventitious and shallow. The leaves are long, narrow, linear, grass- like, light green and arise in a rosette. The flowers have funnel shaped perianth and are fragrant, waxy white, about 25mm long, single or double and borne in a aspike. Stamens are six in number, anthers dorsifixed in the middle. Ovary trilocular, ovules numerous and fruit is a capsule.

Tuberose-Varietal Characters

Tuberose varieties on the basis of number of rows of petals are classified as follows:

Single Flowered Type

- ✰ Varieties having flowers with one row of corolla segments
- ✰ Flowers are extensively used as loose and for extraction of essential oil
- ✰ They are more fragment than double
- ✰ Seed setting percentage is high
- ✰ The flower buds are greenish white
- ✰ Flowers are white
- ✰ Concrete content have been observed to be 0.08 – 0.11 percent
- ✰ Loose flowers of this type of varieties are used for making floral ornaments

Semi-double Flowered Tuberose

- ✰ Flowers with 2-3 rows of corolla segments on straight spikes used for cut flowers
- ✰ Example is cv. Single Double: white flowers with two to three rows of corolla

Double Flowered Type

- ✰ The florets of this flower cultivar have more than three rows of petals, which are white in colour, but tinged with pinkish red
- ✰ It does not bear seeds
- ✰ Spikes are used as cut flowers as well as loose flowers and for the extraction of essential oil
- ✰ Concrete recovery has been found to be 0.0621 per cent
- ✰ Often florets of this double tuberose fail to open completely which results in the reduction of its fragrance

Variegated Type

- ✰ Its florets are similar to Mexican Single except their petals tinged with pinkish colour
- ✰ Golden yellow streaks along the margin of leaves
- ✰ It bears seeds profusely
- ✰ It is highly suited for ornamentation of surroundings
- ✰ Its spikes are the tallest among all popular tuberose cultivars

Propagation

Bulbs are most commonly used for multiplication of tuberose commercially.

Care should be taken in the selection of suitable bulbs. Bulbs free from diseasers, properly matured and having an average diameter of 1.5 cm or above should always be preferred. Tuberose mother bulb measuring 1.5- 2.0 cm diameter produces a clump weighing 200- 250 g having 6-8 daughter bulbs (> 1.5 cm) and 10- 12 bulblets (<1.5 cm). The maximum production of bulbs can be obtained by planting them early. The daughter bulbs should be dug out at next season. Rest period of minimum 4- 6 weeks should be given before planting the bulbs to enhance sprouting percentage and to improve growth, flowering and bulb production. Before planting, bulbs may be treated with Carbendazim (0.1 per cent) or Captan (0.1 per cent) against stem rot or in Hostathion (2 ml/litre) for 30 minutes to minimize root knot nematode attack.

Seed propagation is followed to evolve new varieties. Seeds are difficult to germinate and hence, they are to be sown evenly at the depth of 1.5 cm in well prepared mixture containing leaf mould, garden soil and sand in equal proportions in nursery bed or seed pan. Under favourable soil temperature (26.6°C) and moisture the seeds germinate within two to three weeks time after sowing. When the seedlings are sufficiently grown, they are transferred to the final site.

Cultivation Practices

Climate

Tuberose requires moderate climate with temperature ranging from 20-30°C. Good sunlight is essential for flowering.

Light

Tuberose although not strictly photosensitive, long- day exposure promotes vegetative growth as well as early emergence of the first flower spike and also increases the length of flower spike. A day length of 16 hours promoted growth and flowering

Selection of Site

Tuberose is sensitive to water logging and hence proper drainage should be good. Water logging even for a short period damages the root system and affects growth and flowering.

Soil

Loam and sandy loam soils having a pH in the range of 6.5- 7.5 with good aeration and drainage are suitable. The soil should be rich in organic matter and retain sufficient moisture.

Time of Planting

The time of planting varies in different parts of India. Although tuberose can be planted all round the year, the best planting time in most of the plains is March to May. In and around Coimbatore conditions it is planted during the month of July.

Size of Bulb

Weight and size of bulb influence vegetative growth, flowering and bulb production. Planting of large size bulb (3.0 to 3.5 cm in diameter) significantly improves the yield and quality of flower spike and increases the production of bulbs and bulblets. But in ratoon crop, bulbs with 2.6 to 3.0 cm diameter show better results.

Depth of Planting

Depth of planting depends largely on the size of the bulb. It also varies due to the nature of soil as well as the growing region. Planting depth is more in light soil than that of heavy soil. Based on the above, it may vary from 2.0 to 8.0 cm. In sandy loam soil, planting of bulbs at a depth of 6.0 cm is recommended. Deep planting (6 cm) delays sprouting and reduce bulb production. Shallow planting (2.0 to 4.0 cm) of bulbs helps in more bulb production.

Preparation of Land and Planting

The field should be worked deep to a good tilth and well decomposed farm yard manure at the rate of 5 tonnes/ha should be applied along with 300 kg of neem cake. In beds of convenient size, ridges and furrows should be made with a spacing of 30 cm between two ridges. The bulbs should be planted 30 cm apart half way down the ridge at a depth of 5 cm. Approximately 1, 11, 000 bulbs are required for planting in one hectare as per the spacing recommended.

Fertilizers

The recommended fertilizer dose is as follows (Kg/ha)

Nitrogen	*Phosphorus*	*Potassium*
200 (445 kgs of Urea)	200 (1250 kgs of Super phosphate)	200 (330 kgs of Murate of Potash)

Full dose of phosphorus has to be4 applied at the time of planting while Nitrogen and Potassium have to be applied in three equal split doses *i.e.*, at the time of planting, 30 days after planting and 90 days after planting. However, the above recommendations may be modified according to soil fertility conditions. If a ratoon crop is taken fertilizer has to be applied during the subsequent year.

Growth Regulating Chemicals

Application of CCC at 5,000 ppm or GA3 at 1000 ppm induces early flowering, increases flower stalk production and improves quality of flowers. Foliar spray of ancymidol enhances flowering as well as increases the number of inflorescence. Improvement in bulb yield can be achieved with foliar spray of ancymidol.

Irrigation

Too much moisture in the soil at the time of sprouting results in the rotting of bulbs. After sprouting and establishment irrigation should be given at an interval of 5-7 days based on soil moisture and weather conditions.

Interculture and After Care

After planting the bulbs, the interculture operations begin with the appearance of sprouts through the soil. After sprouting the bulbs, the plants are to be irrigated. Hoeing and earthing up are very important. Hoeing keeps down the weeds and allows aeration in the soil. Earthing up enables the spikes to grow erect despite strong winds and rains. For conserving moisture, maintaining soil temperature and reducing the weed population, mulching with the strips of black polythene, dried grass, chopped straw, farm yard manure or saw dust is very effective. Flowering spikes are to be cut at appropriate maturing or removed at regular intervals.

Weed Control

Weeding is an important operation in tuberose and if neglected it poses a problem to get desired results.Manual weeding is generally practises in tuberose field. Hoeing between plants at regular intervals is helpful in loosening the soil and uprooting the weeds. Hand weeding is laborious, time consuming and expensive. Control of weeds by using chemical is economical, convenient and labour saving compared to hand weeding. Pre plant application of Atrazine at 3.0 kg per hectare causes maximum weed control in tuberose and results in good yield of quality flowers. Tuberose field can be kept almost weed free by the pe-emergence application of Gramaxone at 3.0 kg per hectare causes maximum weed control in tuberose and results in good yield of quality flowers. Tuberose field can be kept almost weed free by the pre-emergence application of Gramaxone at 3.0 l/ ha followed by post- emergence sprays thrice at an interval of 40 days in between the rows of the crop with spraying hood fitted in the nozzle.

Harvesting

In India, tuberose is cultivated for production of flower spikes and loose flowers on a commercial scale. For marketing of flower spikes, the tuberose is harvested by cutting the spikes from the base when quite a good number of flowers open on the spike. The flower spikes are graded according to the stalk length, length of rachis, number of flowers per spike and weight of spike. Straight and strong stems with uniform length and uniform stage of development are preferred. Flowers should be free from bruising injury and from diseases and pests.

Post Harvest Handling

A holding solution consisting of sucrose 2 per cent + $Al(SO_4)_3$ 300 ppm, 4 per cent sucrose + 200 ppm 8- HQC, 4 per cent sucrose + 200 ppm BA was found best for increasing post harvest life and quality of cut spikes of tuberose.

References

Ganesh, S. 2010. Influence of growth regulators and micronutrients on growth, yield and quality of tuberose (*Polianthes tuberosa* l.) cv. Prajwal. M.Sc. (Hort.) Thesis. Tamil Nadu Agricultural University, Coimbatore.

Gudi, G. 2006. Evaluation of tuberose varieties. Thesis submitted to University of Agricultural Sciences, Dharwad, Karnataka.

Gurav, S.B., S.M. Katwate, B.R. Singh, D.S. Kahade, A.V. Dhane and R.N. Sabale, 2005. Quantitative genetic studies in tuberose. *Journal of Ornamental Horticulture,* 8(2): 124-127.

Meenakshi Srinivas, S., Niranjan Murthy and J.L. Karihaloo. 1995. Tuberose hybrids Shringar and Suvasini. *Indian Hort.,* 40 (3): 5-7.

Niranjanmurthy and S. Meenakshi Srinivas. 1997. Genotypic performance and character association studies in tuberose (*Polianthes tuberosa* L.). *J. of ornamental Hort.,* 5(1-2): 31-34.

Patil, V.S., P.M. Munikrishnappa and S. Tirakannanavar, 2009. Performance of growth and yield of different genotypes of tuberose under transitional tract of north Karnataka. *Journal of Ecobiology,* 24(4): 327-333.

Sadhu, M.K. and Bose, T.K. 1973. Tuberose for most artistic garlands. *Indian Hort.,* 18(3): 17-20.

Chapter 17

Agro-technology for Commercial Cultivation of Gerbera in Naturally Ventilated Polyhouse: A Floricultural Activity of Botanic Garden

Satish Kumar and R.K. Roy

Botanic Garden Division, CSIR-National Botanical Research Institute, Lucknow – 226 001, Uttar Pradesh
E-mail: satish_agron@yahoo.co.in

Abstract

Gerbera have wide range of flower colours - yellow, orange, cream-white, pink, brick red, scarlet, maroon, terracotta and various other intermediate shades. The double varieties sometimes have bicolored flowers, which are very attractive. In gerbera, availability of wide range of exotic varieties and their adaptability to grow on wide range of climate makes it a profitable cut- flower crop for farmers. Its cut-flowers stay fresh for 7-10 days and are used for vase decoration as well as in bouquets. NBRI has set up a 'Pilot Facility' for popularization of its cultivation in north Indian plains under Polyhouse conditions in 560 sq. m. area, besides identification of varieties viz., Danaellen, Goliath, Rosalin, Salvadore, Silvester, Sunway and Zingaro suitable for cultivation under the climatic conditions at Lucknow. Economics of this facility has been worked out by taking into account the cost of construction and total income in three years. The estimated net profit is Rs.7 lakhs in 3 years if properly marketed at a sale price of Rs. 5 per flower. This facility was shown to 200 farmers/entrepreneurs/hort. officers/VLFs, others during the report period and was informed about the technical specifications of the polyhouse besides varieties of Gerbera, marketing and economics. This facility drew lot of attention simply due to low cost technology and simple maintenance which progressive farmers/a group of farmers can afford.

Keywords: *Agro-technology, Cultivation, Cut-flower, Gerbera and polyhouse.*

Introduction

Gerbera (*Gerbera jamessonii* Bolus ex Hooker F.), commonly known as 'Transvaal Daisy' or 'African Daisy' is an important commercial flower grown throughout the world in a wide range of soil and climatic conditions. It belongs to the family *Asteraceae* having single and double flowers. Plants are stem less and tender perennial herbs, leave radical, petiole, lanceolate, deeply lobed, sometimes leathery, narrower at the base and wider at toe and are arranged in a rosette at the base. Gerbera is grown for its attractive flowers. The major importance of gerbera in the international flower market is as cut-flower and it suits a wide range of floral arrangements. From commercial point of view, gerbera as a cut-flower ranks sixth, among the ornamental flowers in the world. Nowadays, it has become one of the most important commercial flower crops both for domestic and international markets.

Establishment of Polyhouse

Site Selection

The land should be levelled and availability of good quality water. No wind breaks or multi storied structures present upto 30 meters. Good connectivity to nearest market. Electricity at the site and no high tension electricity wire upto 5 meters.

Soil Structure

Soil pH should be in between 5.5 to 6.5 or it should be maintained at this level to get maximum efficiency in absorption of nutrients. The salinity level of soil should not be more than 1 mS/cm. The soil should be highly porous and well drained to haw better root growth and better penetration of roots.

Soil Sample

For existing plot, it is necessary to take samples from 5 to 8 spots depending on the area. Remove upper 5 cm layer of soil from these spots. Make a v-shaped hole of 6″ depth and scrap soil from slope of hole and this will be a sample from that spot. Similarly take samples from all spots and mix together. Then air dry these sample under shade and then take some sample as a representative sample of the plot.

Soil Sterilization

Before plantation of Gerbera, disinfections of soil are absolutely necessary. In particular, the fungus *Phytopthora* is a menace to Gerbera, The various methods of sterilization are:

1. Sun Sterilization

Cover the soil with plastic for 6-8 weeks. Sun rays will heat up the soil, which will kill most fungai.

2. Chemical Sterilization

a) Hydrogen Peroxide (H_2O_2) with Silver

Wet the beds with irrigation water of neutral pH and EC less than 0.5 mS/cm. In irrigation water; mix hydrogen peroxide with silver at the rate of 35 ml per liter of water (3.5 per cent solution). Use 1.0 liter water for 1.0 m^2 area. Apply this solution uniformly over moist beds using the spout. No need to cover the soil. Plantation can be carried out 4 to 6 hours after fumigation.

b) Formaldehyde

This is reasonably effective on some fungal diseases, but relatively ineffective against many pests. It is used at a strength of 1 part formalin (38-40 per cent) to 50 parts water (approximately 0.5 litre in 25 litres) and the soil is thoroughly drenched (approximately 25 litres being required per m^2) to 25 cm depth. There is a wait period of 20-40 days according to temperature (which must be sufficiently high otherwise the formaldehyde will become polymerized) before the soil can be used.

c) Basamid (Dazomet Granular)

Basamid (Dazomet Granular) is probably the most widely used as it is in a convenient dust-free prill (a coarse powder) in 5 kg packs. It is used at the rate of 30 - 40 gm/m^2. It must be thoroughly mixed with the top 20 cm of damp soil, ideally by rotary cultivator or in heaps of damp soil or growing media spread out on a clean surface in shallow layers. Soil or compost ingredients should be moderately moist and at a temperature of around 9.5°C (49-50°F).

d) Methyl Bromide

It is used at the rate of 25-30 gm/m^2. Methyl bromide has found favour in recent years because of its effectiveness on pests and diseases and also its rapid dissipation.

Bed Preparation

In general, Gerberas are grown on raised beds to assist in easier movement and better drainage. The dimensions of the bed should be as follows:

Bed height: 1.5 ft. (0.45 m)

Bed width: 2.0 ft. (0.60 m)

Pathways between beds: 1.0 ft. (0.30 m)

Organic manure is recommended to improve soil texture and to provide nutrition gradually. At the time of bed preparation (after fumigation) neemcake (@1 kg/m^2) is added as prevention against nematode. All material should be mixed thoroughly for optimum results. The composition of bed material should be such that it should be highly porous, well drained and provide proper aeration to the root system.

Table 17.1: Bed Material Composition According Soil

Material	*Clay Soil*	*Silty Loam Soil*
Red soil	55 per cent	60 per cent
Sand	15 per cent	10 per cent
FYM	30 per cent	30 per cent
Rice husk	4 kg/m^2	2.5 kg/m^2

Table 17.2: Basal Fertilizer Dose (After bed preparation)

Area	*Chemical*	*Quantity*
10 m^2	Single Super Phosphate	2.50 kg
10 m^2	Magnesium Sulphate	0.50 kg
10 m^2	Biozyme Granules	200 gm
10 m^2	Humiguard Granules	200 gm

Details of Polyhouse

Polyhouse Specification

Height: 5 to 6.5 m; Length: North – South; Gutter direction: North – South; Polythene thickness: 200 microns; Vent opening: Along the wind direction. Distance between two adjoining poly houses should be minimum 4 m.

Polyhouse Side Net Opening and Closing

Sufficient ventilation space is required on sides. To protect the plants from the rains, without affecting the air circulation, side curtains should be kept open in slanting position. Open side curtains from 7 am to 6 pm in summer and rainy seasons, whereas open from 9 am to 5 pm in winter, to facilitate maximum air circulation.

Polyhouse Top Shade Net Opening and Closing

During cloudy climate keep open, otherwise close from 10 am to 4 pm. During cold nights close the shade net. In summer close from 9.30 am to 5 pm. Wash top of plastic with clean water monthly. Always maintain hygiene and sanitation inside as well as outside polyhouse. To control light intensity and solar radiation, white shade net (50 per cent) is used. Approximately 35,000 to 40,000 lux light intensity is required on the plant level.

Planting

While planting Gerbera plants, the crown of plants should be 1 - 2 cm above soil level. As the root system establishes; the plants are pulled down. Therefore, the crown must be above the ground level at planting and also throughout the life cycle. Plant the seedlings without disturbing the root ball (25 per cent above the soil: 75 per cent below the soil). Generally two rows should be planted on one bed at 37.5 cm distance and 30 cm distance between the plants in one row *i.e.*

Figure 17.1. Model Polyhouse for Gerbera Cultivation

- ✰ Row to Row spacing: 37.5 cm
- ✰ Plant to Plant spacing: 30.0 cm

Rake the soil surrounding the plant every fortnight for aeration. After plantation, maintain the humidity at 80 - 90°C for 4 - 6 weeks to avoid desiccation of plants. Avoid excessive watering to gerbera.

Irrigation

Water quality should be as follows: pH- 6.5 - 7.0; EC < 0.7 mS/cm; T.D.S. < 450 ppm; Hardness < 200 ppm. To lower the pH of water, Nitric acid (HNO_3) or Phosphoric acid (H_2PO_4) can be added to the tank. The water requirement of Gerbera plant is approximately 300 to 700 ml per plant per day depending upon the season. In hot summer, loggers can be, used to maintain the humidity of the air. The foggers should not be operated for more than 30 seconds at any given point in time. During summer season, apply water to the edges of the beds frequently by using shower to minimize the evaporation losses and to maintain micro climate. For this purpose provision for water outlets (2.5 cm diameter pipe) should be made inside the polyhouse. The relative humidity of air should not exceed 80 per cent, as it will lead to conditions conducive for the occurrence and survival of fungal pathogens, especially Botrytis, on the flowers.

Fertilization

After three weeks of plantation apply N:P:K 1: 1: 1 (*e.g.* 19:19:19) @ 0.4 gm/ plant every alternate day with EC 1.5 mS/cm for first three months during the vegetative phase to have better foliage. Disbudding operation had to be carried out until 16 -18 fully developed leaves are present on the plant. Thereafter, the generative stage fertigation can be commenced. The generative stage fertigation comprises of N: P: K 2:1:4 (*e.g.* N:P:K 15:8:35) @ 0.4 gm/plant every alternate day with EC 1.5 mS/cm to enhance the quantity and quality of flowers. Micronutrients (*e.g.* Fertilon Combi II, Microsole B, Rexolin, Sequel and Mahabrexil @ 40 gm per 1000 lit of water) should be given daily or weekly as per the deficiency symptoms. Add organic manures with EC less than 2 mS/cm at every 3 months interval to maintain proper C: N Ratio. Always do the detail soil analysis every 2 - 3 months to decide specific nutrient schedule.

Crop Duration, Harvesting and Treatment of Flowers

Gerbera is a 30 - 36 months crop. The first flowers are produced 7 - 8 weeks after plantation when plants are with 14 to 16 leaves. The average yield is 240 flowers per m^2 (6 plants/m^2). The flowers are harvested when 2 - 3 whorls of stamens have entirely been developed: this will decide the vase life of flowers.

Pluck the flowers in the morning or late in the evening or during the day when temperature is low. After harvesting, cut the heel of the flower stalk by giving an angular cut. Place the flowers immediately in 2-3 cm water for four hours at 14-15°C. Always add commercial bleach/Sodium Hypochlorite @ 7-10 ml or Citric acid + Ascorbic acid @ 5 ml each/lit of water. Sleeves the individual flower with polythene bag of size 4.5 × 4.5 cm. Make bundle of 10 flowers. Pack the flowers in a box with following dimensions. Generally 250 to 300 flowers are packed per box. A good gerbera flower stalk length 25-30 cm and diameter of flower is 10-12 cm. A Gerbera cut flower has a minimum vase life of 8-10 days.

Pests

- ☆ **Aphids:** Cause deformed leaves, excrete some substance on which fungus develops.
- ☆ **Caterpillar:** Eat leaves voraciously making circular holes in the leaf lamina. It causes white spots on the petals in case of flower attack.
- ☆ **Cyclamine Mites:** Older leaves are curled up. Younger ones being deformed and leathery, deformed flowers or petals are missing. Inward curling and discolouration of petals.
- ☆ **Greenhouse Whitefly:** It occurs when climate is hot and dry. Feeds on the lower side of leaves, excrete large quantity of honey dew like substance which leads to development of black sooty moulds on the leaves.
- ☆ **Leaf Miner:** White specks on leaves caused by flies. White serpentine tunnels in leaves caused by larvae, which stays in soil.
- ☆ **Mealy Bugs:** Mealy bugs are identified as white cottony insects, feeding on leaves and tender growing points. The feeding damage is characterized by yellowing and distortion of plant parts. During feeding, the insect secretes honey dew, which serves as a medium for black sooty mold fungi.
- ☆ **Red Mites:** Suck the sap from lower sides of the leaves causing development of brown spots on lower surface of leaves resulting in marginal drying of leaves. Webbing on the flower petals.

- **Root knot Nematode:** Yellowing of leaves, stunted growth of the plant with reduced leaves size, knots on roots. Water logged condition in the greenhouse and muddy water during rainy season are favourable conditions for nematode growth and spread.
- **Snails/Slugs:** Snails and slugs thrive under plant debris and soil under cool and moist conditions. They come out to feed during night hours. The symptoms are characterized by circular feeding holes on the leaves and flower petals.
- **Thrips:** Causes white specks or stripes on ray florets; flower heads may be deformed. Silvery, grayish spots on the leaves; Brown spots on leaf petioles/mid vein.

Diseases

- **Alternaria leaf Spot:** Develops when moisture persist on leaf surface for longer duration. Black circular spots appear on leaves.
- **Bacterial Blight:** Yellowish oily spots on the leaves later turn brown. Brown discolouration along the mid vein. Wilting off lower bud and brown spots on the stems.
- **Botrytis:** Occurs especially when the relative humidity of the air is more than 92 per cent for two hours in the morning. Gray spots on the flower petals-rot in the heart of flower.
- **Crown Rot:** Caused by *Phytophthora cryptogea* results in wilting disease of Gerbera, Crown of the plant becomes black.
- **Fungal Complex:** Plant becomes weak and stunted, with poor quality stems. This is a combined infection of *Cylindrocarpon destructans, Fusarium solen* and *Fusarium oxysporum.* If leaf stem is cut, you can see black vessels. Blocking of crown portion with brown discolouration.
- **Phyllody:** This is an irreversible phenomenon, which is the abnormal development of floral parts to leafy structures. It is caused due to combined effect of unfavorable biotic as well as abiotic factors.
- **Powdery Mildew:** White powdery growth on the leaf lamina. In case of severe attack leaves start curling.
- **Root Rot:** Caused by *Pythium*. Initially dropping of younger leaves, finally wilting of the plant. Root skin is easily removed.

Table 17.3: Pest Control in Gerbera

Pests	*Suggested Control Measures*	*Concentration per lit.*
Whitefly	Rogor (Dimethoate)	2.0 ml
	Neemazol	2.0 ml
	Confidor (Imidacloprid)	0.5 ml
	Pride (Acetamiprid)	0.4 gm

Pests	*Suggested Control Measures*	*Concentration per lit.*
Leaf Miner	Nuvan (Dichlorovos)	1.0 ml
	Acephate (Acephate)	1.5 gm
	Cypermethrin	0.5 ml
Thrips	Nuvan (Dichlorvos) + Nuvacron (Monocrotophos)	1.5 ml + 2.0 ml
	Rogor (Dimethoate)	2.0 ml
	Pride (Acetamiprid)	0.4 gm
Red Mites	Wettable Sulphur	1.5 gm
	Kelthane (Dicofol)	1.5 ml
Cyclamen Mites	Wettable Sulphur	1.5 gm
	Karathane (Dinocap)	0.4 ml
Catterpiller	Thimet (Phorate) (S)	2.0 gm/plant
	Decis (Deltamethrin)	0.5 ml
	Larvin (Thiodicarb)	0.4 gm
Nematode	Neem cake	30 to 50 gm/plant
	Metacid (Methyl parathion)	2.0 ml
	Benlate (Benomyl)	3.0 gm
	Nematoguad (Pcealomyces)	5.0 gm
Snail	Snail Kill (Iron EDTA complex)	1 pellet/m^2
	Lannate (Methomyl)	1.5 gm
Mealy Bugs	Nuvan (Dichlorvos) + Nuvacron (Monocrotophos)	1.5 ml + 2 ml

Table 17.4: Disease Control in Gerbera

Disease	*Suggested Control Measures*	*Concentration per lit.*
Root Rot	Aliette (Fosetyl Alluminiurn) (SP)	1.0 gm
	Bavistin (Carbendazirn) (D)	2.0 gm
	Captaf (Captan) (D)	2.0 gm
Crown Rot	Aliette (Fosetyl Alluminiurn) (SP)	1.0 gm
	Kocide (Copper hydroxide) (D)	2.0 gm
Fungal Complex	Bavistin (Carbendazirn) (D)	2.0 gm
	Hydrogen Peroxide with Silver (D)	3.0 ml
	Streptocyclin + COH (D)	0.2 gm + 1.5 gm
Alternaria Leaf Spot	Dithane M-45 (Mancozeb) (SP)	1.5 gm
Powdery Mildew	Wettable Sulphur (SP)	1.5 gm
	Index (Myclobutanil) (SP)	0.5 gm
	Rubigan (Fenremol) (SP)	1.0 ml
Botrytis	Dithane M-45 (Mancozeb) (SP)	1.5 gm
	Index (Myclobutanil) (SP)	0.5 gm

Disease	Suggested Control Measures	Concentration per lit.
Bacterial Blight	Streptocyclin (SP)	0.2 ml
	Blitox (Copper oxychloride) (D)	1.5 gm
	Kocide (Copper hydroxide) (D)	2.0 gm
Phyllody	Streptocyclin + COC (D)	0.2 gm + 1.5 gm
	Hydrogen Peroxide with Silver (D)	3.0 ml

D: Drench @ 50 - 100 ml/plant; SP: Spraying.

Spraying of Pesticides, Fungicides and Fertilizers for Better Results in Gerbera

pH of the solution should be between 6 to 6.5. It is maintained by acid treatment to water 12 hours before spray. Use solution immediately after preparation. Do not store mixture for more than 2 hours after preparation. Smaller droplet size helps to cover maximum leaf surface area.

Flower Bent of Gerbera

Loss of cell turgidity and under nutrition (lack of Calcium).

- ☆ **Pre-harvest stem break:** High root pressure and high humidity in the air.
- ☆ **Premature wilting of flower:** Cloudy weather followed by bright sun or carbohydrate depletion.
- ☆ **Double-faced flower:** A physiological disorder caused by imbalance of nutrients. Too much growth too little flower buds.
- ☆ **Non-uniform flower blooming:** Physical injury to flower stem/pest damage/phytotoxicity.
- ☆ **Short stem length:** High salinity level, moisture stress, low soil temp.
- ☆ **Hollow stem:** Inadequate sunlight during cloudy weather, leading to lack of carbohydrate accumulation in the flower stalk.
- ☆ **Petal drop:** High turgor pressure leads to the petals dropping off with the slightest touch at the time of packing in sleeves.

Deficiency Symptoms of different Nutrients in Gerbera

- ☆ **Nitrogen:** General yellowing starts on older leaves and then moves gradually upward because nitrogen is translocated out of older leaves to the new growth under deficiency.
- ☆ **Phosphorus:** Brownish to purplish discolouration along the veins or the underside of the older leaves.
- ☆ **Potassium:** Marginal necrosis of old leaves.
- ☆ **Calcium:** Extreme yellowing of young leaves.
- ☆ **Magnesium:** Interveinal chlorosis on older leaves, leaves get thick and crispy.

- ✰ **Iron:** Interveinal chlorosis on young leaves. Serious deficiency results in a yellowish-white coloring.
- ✰ **Zinc:** Chlorosis, one half of leaf blade ceases to expand and develop while other half is normal *i.e.* C shaped leaf structure.
- ✰ **Manganese:** Leaves turn yellowish, starting with younger ones; veins remain green, heavy chlorosis.
- ✰ **Copper:** Chlorosis in younger leaves; flower develops bad.
- ✰ **Molybdenum:** Chlorosis on the edges of leaves.
- ✰ **Boron:** Bases of younger leaves are black colored.

Control Measure

Chelated sources (like Microsole, Tracel, Micnelf) of these microelements as a foliar spray.

Projected Economics of Gerbera (in three years)

Area : 560 Sq. m. (28 m × 20 m)

Spacing : 0.30×× 0.30 m

Total no. of plants : 3000 nos.

Total cost : Rs. 8.0 Lakh

(Rs. 6.0 Lakh for polyhouse and planning material + Rs. 2.0 Lakh for maintenance)

Production of flowers : 3.0 Lakh flowers (Av. 40/flowers/plant/year)

Figure 17.2. Inner view of Gerbera Polyhouse, CSIR-NBRI Botanic Garden, Lucknow.

Value of flowers : Rs 15.0 Lakh (Rs. 5.0 per flower)

Net profit : Rs. 7.0 Lakh

Conclusion

The technology for the production of cut-flowers in the naturally ventilated polyhouse was standardized for the plains of northern India and transferred to the progressive farmers as a part of ongoing promotional activities of floriculture.

Further Readings

Ahlawat, T.R., Barad, A.V. and Jat Giriraj.2012. Evaluation of gerbera cultivars under naturally ventilated poly house. *India J Hort.*, 69: (4) 606-608

Chobe, R.R., Pachankar, P.B. and Wanade, S.D.2010. Performance of different cultivars of gerbera under poly house condition. *The Asian J Hort.*, 2: 333-335.

Choudhary, M.L. and Prasad, K.V. 2000. Protected cultivation of ornamental crops-an insight. *Indian Hort.*, 45(1): 49-53.

Kumar, S., Roy, R.K., Tewari, S.K., and Tuli, R. 2010. Gerbera make way to Lucknow an initiative by NBRI. *Floriculture Today*. 14(9):10-12.

Chapter 18

Effects of different Growing Media on Growth and Yield of Anthurium (*Anthurium andreanum* Var. 'Avo-Victoria') for Cut Flower Production under Polyhouse

Tapas Kumar Chowdhuri

PROVIDE ADDRESS
Agri-Horticultural Society, Alipore Road, Kolkata, West Bengal.
E-mail: tkc.hort@gmail.com

Abstract

An experiment was carried out in the "The Agri-horticultural Society of India" with an aiming to standardization of growing media for cut flower production of Anthurium (Anthurium andreanum Var. 'Avo-Victoria') under polyhouse. There were eleven treatments combinations (T_1:Soil, $T_{2:}$ Soil + OLM, $T_{3:}$ Soil + CM, $T_{4:}$ Soil + LM, $T_{5:}$ Soil + OLM + CM, $T_{6:}$ Soil + OLM + LM, $T_{7:}$ Soil + OLM + CM + LM, $T_{8:}$ Soil + OLM + CM + LM + BM, $T_{9:}$ Soil + OLM + CM + LM + SM, $T_{10:}$ Soil + OLM + CM + LM + NS and $T_{11:}$ Soil + OLM + CM + LM + BM), where basic growing media was soil and enriched by adding old lime mortar, decomposed cow dung manure, leaf mould, bone meal, stera meal, neem shield and garden bloom. It was observed from this study that treatment combination of T_{11} showed better performance in terms of vegetative growth as well as flowering behavior significantly. Others treatments like T_7, $T_{8,}$ T_{10} are also found very positive response in cut flower production. All these four treatments are effectively increased NPK contents in leaves of anthurium, but it has been found that leaf contents of N and K was not related with number of flower production, while P content showed positive and significant (P=0.01) relationship with flower yield.

Keywords: *Anthurium, Polyhouse, Growing media, Growth, Flower yield.*

Introduction

Flower production in polyhouse is now widely practiced throughout the world. However, in our country this is relatively new, and little work has been done to standardize this technique for commercial exploitation. Anthurium (*Anthrium andreanum* Var. 'Avo-Victoria') is one of the most important commercial flowering crops growing all over world as a cut flower and now it being popularized in India. Unfortunately, Indian standard for production of quality flowers is not up to the mark as required by the world trade market. There are so many genetically modified improved varieties are released and it is available in the market, but suitable growing media for proper growth and development of these varieties are not standardized. Very limited research work has been done in this aspect in India. In the light of the above, the present study was undertaken in a poly house with emphasis on standardization of growing media for cut flower production of anthurium.

Highly organic, well aerated medium with good water retention capacity and drainage is generally used for growing anthurium. Matysiak and Nowak (1996) reported that *Anthurium cultorum* under natural condition grows better in peat: perlite (3:1) substrate. Dufour and Clairon (1997) noticed that 2:1 mixture of volcanic ash and wood chips as best for cut flower production of anthurium. Rajeevan *et al.* (2002) reported from TNAU (Coimbatore) that *Anthurium andraenum* cv. 'Temtation' under 75 per cent shade net produced good quality and quantity of flower as well as suckers in leaf mould in combination of coco peat. In an another experiment, Pawar *et al.* (2002) tried various substrates (coconut coir pieces, brick pieces and wood charcoal alone or in combinations); they found largest flower size (63.25 sq.cm) in coconut coir pieces + wood charcoal and smallest in brick pieces (32.02 sq.cm), while the longest flower stalk (41.44 cm) was recorded in wood charcoal and the shortest (26.6 cm) in brick pieces.

Materials and Methods

The investigation was carried out under polyhouse at The Agri-Horticultural Society of India and department of Horticulture, Institute of Agricultural Science, Calcutta University. The sub-tropical climatic condition comprised of average temperature (max.21.7-39.62°C and mini. 15.13-26.49°C), relative humidity (max.95.54-78.41 per cent and mini.41.24-78.03 per cent), light intensity(3500-4000 lux) and annual rainfall (1200-1400 mm). Poly-house having facilities with Fan-pad cooling system, additional shade net (50 per cent), garnet, foggers, drip and sprinkler irrigation system. Plants were grown initially in 10cm diameter earthen pots for one month, when growing media was used only loam soil. Then it was transplanted in 25 cm earthen pots, when basic growing media was soil and enriched by using of Old lime mortar, Cow dung manure, Leaf mould, Bone meal, Stera meal, Neem shield and Garden bloom. Tissue culture plant materials at four leaf stages with an average leaf length of 4-5cm were used. All pots were reported every year by removing ½ part of the total growing medium and refilling with the same, but fresh medium. During reporting all ineffective suckers and some roots were removed. There were eleven combinations of different ingredients used in this experiment under mentioned with four replications, each treatment having 20 nos. of potted plants.

Table 18.1: Treatments Details

Name of Treatments	Details of Treatments	Name of Treatments	Details of Treatments
T_1	Soil	T_6	Soil + OLM + LM (3:1:1)
T_2	Soil + OLM (4:1)	T_7	Soil + OLM + CM + LM (2:1:1:1)
T_3	Soil + CM (4:1)	T_8	Soil + OLM + CM + LM + BM(2:1:1:1 + 50g)
T_4	Soil + LM (4:1)	T_9	Soil + OLM + CM + LM + SM(2:1:1:1 + 50g)
T_5	Soil + OLM + CM (3:1:1)	T_{10}	Soil + OLM + CM + LM + NS(2:1:1:1 + 50g)
		T_{11}	Soil + OLM + CM + LM + GB(2:1:1:1 + 50g)

OLM: Old lime mortar; CM: Cow dung manure; LM: Leaf mould; BM: Bone meal; NS: Neem shield; SM: Stera meal; GB: Garden bloom.

Observation was taken at regular interval and data was recorded based on plant height (cm), no. of leaves/plant, No. of suckers/plant, days required for flowering, no. of flowers/plant/year, spathe length(cm), spathe width(cm), stalk length(cm), longevity of flower on plant(days) and vase life of flower in tap water(days). Chemical analysis was done for estimation of minerals contains in the leaf like N (Kjeldahl method), P (Vanado-molybdophosphoric yellow colour method) and K (Perkin flame photometer). The statistical analysis of the data was carried out following Fisher's analysis of Variance Technique as described by Gomez and Gomez (1984).

Table 18.2: Effect of different Growing Median on Vegetative Growth of Anthurium

Treatments	Plant Height (cm)	No. of Leaves/ Plant	Leaf Area (sq.cm)	No. of Suckers/ Plant/Year
T_1	24.50	4.80	90.80	2.00
T_2	26.10	5.40	114.30	2.40
T_3	26.30	6.00	119.40	2.80
T_4	27.00	5.80	112.40	2.80
T_5	28.10	6.30	129.60	3.10
T_6	28.70	6.10	117.40	3.10
T_7	30.90	6.80	129.60	3.50
T_8	33.60	7.20	200.70	4.10
T_9	24.50	4.00	112.40	2.50
T_{10}	27.50	6.00	129.00	2.70
T_{11}	36.40	9.00	254.60	4.70
LSD (*P=0.05*)	1.24	0.38	4.03	0.19

Treatments details in Table 18.1.

Results and Discussion

Data in Table 18.2 clearly showed the effect various growing media on growth parameters like plant height, number of leaves per plant, leaf area, and no. of suckers produced annually by anthurium. The maximum plant height(36.4cm), number of leaves per plant(9.0), leaf area(254.6 sq.cm) and number of suckers(4.7) per plant were recorded under $T_{11:}$ Soil + OLM + CM + LM + GB, whereas the minimum plant height was recorded in $T_{1:}$ Soil and $T_{9:}$ Soil + OLM + CM + LM + SM). The next best treatment appeared to be $T_{8:}$ Soil + OLM + CM + LM + BM followed by $T_{7:}$ Soil + OLM + CM + LM).

The potting mixture showed pronounced effect on flowering of anthurium (Table 18.2). Earliest flowering was recorded when anthurium plants were grown in a potting media containing $T_{10:}$ Soil + OLM + CM + LM + NS closely followed by $T_{9:}$ Soil + OLM + CM + LM + SM), whereas flowering was advanced by 66 and 59 days respectively, when compared with soil (T_1). Others treatments like $T_{11,}$ T_8 and T_7 also advanced flowering to noticeable extent. In contrast, flowering was slightly delayed (by 14 days) in a medium containing (T_2) soil + OLM. From Table 18.3 also revealed that the treatments had marked influence on flower production and flower quality of anthurium. Thus, T_{11} brought about (Figures 18.1 and 18.2) maximum number of flowers/plant/year (8.5), spathe length (10.2 cm), spathe width (6.1cm), stalk length (30.8 cm) and even longevity of flowers on plant (18.4days) and vase life of flower in tap water (11.4 days).

Table 18.3: Effect of different Growing Media on Flowering of Anthurium

Treatments	Days to Flowering	No. of Flowers/ Plant/Year	Spathe Length (cm)	Spathe Width (cm)	Stalk Length (cm)	Longevity of Flower on Plant (Days)	Vase Life of Flower in Tap Water (Days)
T_1	351	2.50	7.10	5.00	20.10	15.60	9.40
T_2	365	3.10	7.30	5.30	20.40	17.60	10.80
T_3	339	3.60	7.70	5.80	20.40	16.40	10.40
T_4	356	3.40	7.50	5.70	20.60	14.20	8.30
T_5	345	4.10	7.90	5.70	16.00	17.50	10.50
T_6	351	3.80	7.80	5.70	20.40	16.30	9.50
T_7	317	4.40	8.50	5.90	21.20	16.30	9.10
T_8	300	6.10	9.70	5.90	27.30	17.80	10.40
T_9	292	1.90	8.10	5.10	20.20	13.80	8.10
T_{10}	285	4.70	9.10	5.70	25.90	17.60	9.50
T_{11}	300	8.50	10.20	6.10	30.80	18.40	11.40
LSD (*P=0.05*)	2.33	0.27	0.26	0.16	0.68	0.50	1.17

Treatments details in Table 18.1.

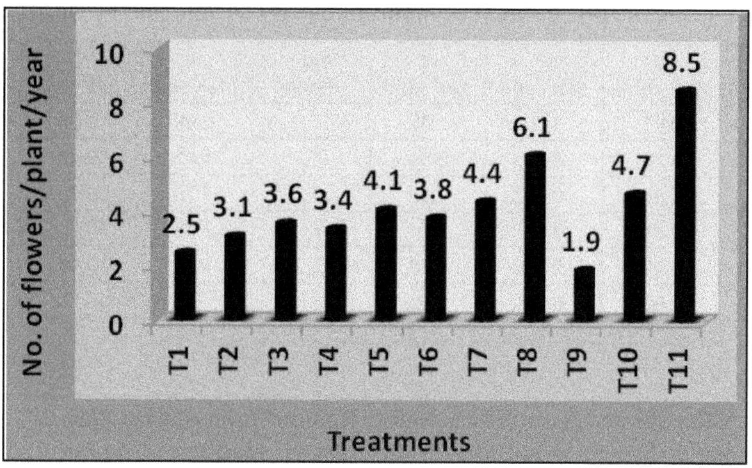

Figure 18.1: Effect of different Growing Media on Flower Production of Anthurium.

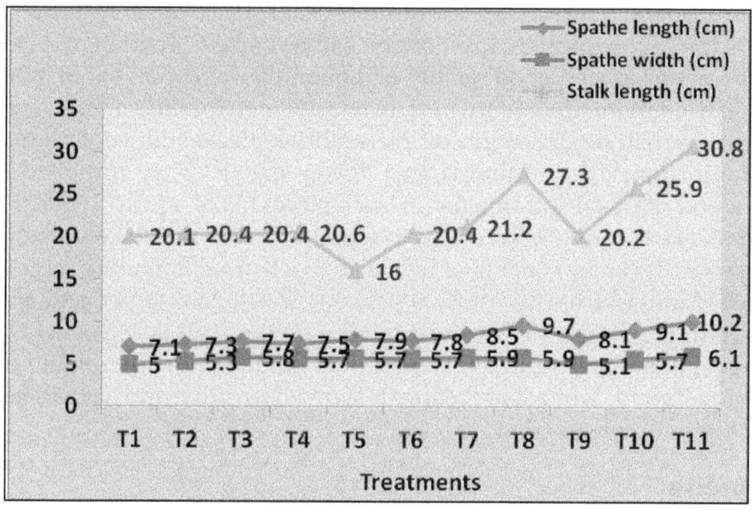

Figure 18.2: Effect of different Growing Media on Improvement of Flower Quality of Anthurium.

Treatments $T_{7,}$ $T_{8,}$ T_9 also showed beneficial effect on flower production and its quality, although the magnitude of increase by these treatments was much less pronounced as compared to $T_{11.}$ At the time of flower emergence, the leaves were collected from plants under all treatments and analyzed in the laboratory for NPK contents as affected by various potting mixtures (Table-4). In anthurium, the NPK contents of leaves varied widely depending on treatments (0.67-1.76 per cent N, 0.20-0.48 per cent P and 1.56-6.84 per cent K). Although the highest N and K contents were recorded in the treatment T_9 plants, showing leaf N, P and K content of 1.20 per cent, 0.46 per cent and 5.62 per cent, respectively, T_{11} showed best result of growth and flowering. Soil grown plants had minimum leaf of N, P and K contents.

Table 18.4: Level of NPK Content of Anthurium Plant as Affected by Growing Media

Treat-ments	N (per cent)	P (per cent)	K (per cent)	Treat-ments	N (per cent)	P (per cent)	K (per cent)
T_1	0.67	0.20	1.56	T_7	0.98	0.36	3.02
T_2	0.78	0.28	2.56	T_8	1.17	0.48	5.92
T_3	0.89	0.32	1.82	T_9	1.76	0.32	6.84
T_4	0.86	0.30	1.74	T_{10}	1.45	0.26	6.24
T_5	0.92	0.33	2.82	T_{11}	1.20	0.46	5.62
T_6	0.92	0.33	2.62				

From the above result, it was found that old lime mortar played a vital role for production flower of anthurium in terms of quality and quantity along with cow dung manure, leaf mould and bone meal or garden bloom, because it improve calcium content in the soil, improve aeration in the soil and increase moisture content in the soil, which is relevant to this plant. Rajeevan *et al.* (2002) reported from TNAU (Coimbatore) that *Anthurium andraenum* cv. 'Temtation' under 75 per cent shade net produced good quality and quantity of flower as well as suckers in leaf mould in combination of coco peat. In an another experiment, Pawar *et al.* (2002) tried various substrates (coconut coir pieces, brick pieces and wood charcoal alone or in combinations); they found largest flower size (63.25 sq.cm) in coconut coir pieces + wood charcoal and smallest in brick pieces (32.02 sq.cm), while the longest flower stalk (41.44 cm) was recorded in wood charcoal and the shortest (26.6 cm) in brick pieces. It can be concluded here also that during flowering stage if the leaf content of N ranges from 1.17 to 1.20 per cent, P 0.46 to 0.48 per cent and K 5.62-5.92 per cent, maximum yield of best quality flowers can be expected. The study further revel that leaf N or leaf K content was not related with number of flowers produced per plant, while leaf P content showed positive and significant($P=0.05$) relationship with flower yield.

Conclusion

For Anthurium growing under polyhouse in sub-tropical zone for cut flower production, growing media can be recommended soil along with well decomposed cow dung manure, leaf mould, old lime mortar(2:1:1:1) and enriched with garden bloom or bone meal @ 50g/plant/year.

References

Dufour, L. and Clairon, M. (1997). Advances in fertilization of *Anthurium hybrid* in Guadeloupe. *Acta Horticulture*, 450: 433-438..

Matysiak, B. and Nowak, J. (1996). Effect of substrate on the growth of *Anthurium cultorum* and *Spathiphyllum wallisii*. *Zeszyty Problemowe Postepow Nauk*, 429: 241-244.

Pawar, G. M.; Patil, M. T. and Gaikwad, A. M. 2002. Effect of substrate on anthurium culture. *Indian Society of Ornamental Horticulture*, pp.326-327.

Rajeevan, P. K., Valsalakumari, P. K., Geetha, C. K.; Ravidas, L.; Kumar, V. and Bhattacherjee, S. K. 2002. All India Coordinated Research Project on anthurium. *Technical Bulletin*, AICRP, No. 18: 25-26.

Chapter 19

Studies on Growth and Cut Flowers Production of Gerbera (*Gerbera jamesonii*) under Naturally Ventilated Polyhouse

Satish Kumar and R.K. Roy

Botanic Garden Division, CSIR-National Botanical Research Institute, Lucknow – 226 001, Uttar Pradesh
E-mail: satish_agron@yahoo.co.in

Abstract

The present investigation was carried out to study the performance of gerbera varieties growing in naturally ventilated polyhouse in Lucknow, Uttar Pradesh. Seven varieties viz., Salvadore, Silvester, Goliath, Zingaro, Sunway, Dana Ellen and Rosaline were planted on growing media viz., soil and sand: FYM: rice husk having pH 6.2. Planting was done in beds (26.0×0.75×0.60 m) in two rows (0.30×0.30 m) at a distance of 0.30 m between plants. Observations on growth and flower characteristics were recorded periodically for a period of one year and statistically analyzed. The results revealed that the variety Rosaline recorded maximum plant height (33.67 cm), number of leaf/plant (31.33), leaf length (21.67 cm), leaf breadth (10.33 cm) and number of suckers/plant (11) followed by Dana Ellen and Silvester. Rosaline recorded minimum days (45 days) for bud initiation and days taken to first flower since planting (52.67 days) followed by Dana Ellen. Highest values for stalk length (61 cm), flower stalk diameter (1.99 cm), flower diameter (11.63 cm) were recorded in Rosaline. On the basis of the investigation, it was concluded that all the seven varieties had variation in respect of growth and flowering under naturally ventilated polyhouse under standard growth and climatic conditions. Three varieties Rosaline, Dana Ellen and Silvester were recommended for commercial cultivation of gerbera for the production of cut-flowers in and around Lucknow, Uttar Pradesh.

Keywords: *Cut-flowers, Gerbera, Growth, Naturally ventilated polyhouse and Production.*

Introduction

Gerbera as a cut-flower has high demand in domestic and international markets. Due to globalization in different parts of the world, per capita consumption of flower in most countries is increasing rapidly. In recent years, commercial production of gerbera has become a major venture in India. The production technology for viable production needs to be standardized for Indian climate and soil. Moreover, the commercial floriculture venture is very much cultivar specific. Considering the importance of this crop, there is a prime need for its improvement through selection of suitable varieties based on vegetative and floral characteristics. The yield varies due to environmental conditions and genetic potentialities of the varieties. All varieties may not be adaptable for cut-flower production in all agro-climatic zones of Uttar Pradesh. However, the genetic diversity that exists in the crop would greatly facilitate selection of suitable types for transitional tracts of Uttar Pradesh.

Method and Materials

Experimental Sites

The experiment was conducted during the year 2009-10 at Botanic Garden, CSIR-National Botanical Research Institute, Lucknow under naturally ventilated polyhouse condition. The polyhouse was constructed with UV stabilized polyfilm (200 μ thickness) as cladding material. The centre height and height below gutter of the polyhouse were six and four meters, respectively. A shade net (white colour) with 50 per cent shade was placed below the roof at four meter height from ground level with a provision to spread and fold depending on the light transmission. The sides of polyhouse were covered with 50 per cent shade net for one meter height (white colour) in order to get proper ventilation and on top kept 0.8 m height for effective ventilation. Soil bed was provided with two rows of drip line at 0.30 m distance and inline drippers at 0.30 m distance with a discharge rate of 1.3 litres per hour. Water soluble fertilizers were dissolved in separate fertigation tank with provision to pump along with water to irrigate the plants.

Crop Management

Seven varieties *viz.*, 'Salvadore', 'Silvester', 'Goliath', 'Zingaro', 'Sunway', 'Dana Ellen' and 'Rosaline' were planted on growing media *viz.*, soil and sand: FYM: rice husk having pH 6.2. The experiment was laid out in randomized block design with three replications. The beds of 0.75 m width were formed along the width of polyhouse to accommodate two rows of gerbera on each bed. Tissue culture raised plants with 4-6 leaves were planted at a spacing of 0.30×0.30 m on beds. Before preparation of beds, mixture of fine sand @ 1 cft/m^2 and decomposed dry FYM @ 5 kg/m^2 as well as NPK @ 20:10:20 g/m^2 were applied as basal dose. The air temperature inside the polyhouse was set at 22°C and maintained with the help of fogger controlled cooling system. Fertigation was done through drippers near the root zone on daily basis and the plants were allowed to put forth 20-25 leaves per plant by periodical removal of flower buds during initial three months for maintenance of good health. Initial doses of NPK grade 19:19:19 were applied through drip at one week interval for three months. After the required foliage was established, the flower buds were allowed for cut-flower production. The application of fertilizers and water levels were changed from the vegetative phase to reproductive phase and it was applied with 50:25:50 NPK at 15 days interval through drip.

Figure 19.1. Silvester, Rosaline, Dana Ellen (left to right)

Plant Sampling and Analysis

Fifteen plants per replication were randomly tagged for growth and floral observation. Following characteristics were considered for data recording: Plant height (cm), number of leaves/plant, leaf length (cm), leaf breadth (cm), number of suckers/plant, bud initiation (in days), days taken to first flower since planting, stalk length (cm), flower stalk diameter (cm) and flower diameter (cm).

Data Analysis

Observations on growth and flower characteristics were recorded periodically for one year and statistically analyzed using Randomized Block Design to draw conclusion.

Results and Discussion

The results obtained from the present investigation on various parameters exhibited significant differences among the varieties as is evident from Table 19.1. Variety Rosaline recorded maximum plant height (33.67 cm), number of leaf/plant (31.33), leaf length (21.67 cm), leaf breadth (10.33 cm) and number of suckers/ plant (11) followed by 'Dana Ellen' and 'Silvester', whereas, the lowest plant height (24.33 cm), number of leaf/plant (24.33), leaf length (15.33 cm), leaf breadth (7.67 cm) and number of suckers/plant (9) recorded in Goliath variety. The differences in various vegetative growth parameters might be attributed to inherent genetic characters of the varieties evaluated. The same inference has been reported by Singh and Ramachandran (2002), Biradur and Khan (1996), Kumar and Yadav (2005), and Praneetha (2006).

'Rosaline' recorded minimum days (45 days) for bud initiation and days taken to first flower since planting (52.67 days) followed by 'Dana Ellen'. Highest values for stalk length (61 cm), flower stalk diameter (1.99 cm), flower diameter (11.63 cm) were recorded in Rosaline. With respect to flower characters, all the varieties recorded significant differences for flower characters. The important character *viz.*, flower stalk length was highest in Rosaline followed by Dana Ellen The same was reported by Barooah *et.al.* (2009), Barreto and Jagtap (2006), Thangam *et.el.*,(2009).

Table 19.1: Comparative Analysis of Vegetative Growth and Flower Characteristics of Gerbera Varieties

Varieties	Vegetative Growth Characteristics					Flower Characteristics				
	Plant Height (cm)	Number of Peaf/Plant	Leaf Length (cm)	Leaf Breadth (cm)	Number of Suckers/ Plant	Bud Initiation (in days)	Days taken to First Flower since Planting	Stalk Length (cm)	Flower Stalk Diameter (cm)	Flower Diameter (cm)
'Salvadore'	28.33	26.67	18.33	8.67	10.00	49.00	56.33	59.33	1.76	10.03
'Silvester'	30.33	29.67	18.33	9.00	10.00	48.33	55.00	59.33	1.80	10.05
'Goliath'	24.33	24.33	15.33	7.67	9.00	56.33	63.33	57.33	1.65	9.90
'Zingaro'	25.33	25.33	15.67	8.00	9.00	54.67	62.67	58.33	1.70	9.86
'Sunway'	27.33	26.67	17.33	8.33	9.33	49.33	57.67	59.00	1.74	9.93
'Dana Ellen'	31.67	30.33	19.67	9.33	10.67	48.00	54.67	59.67	1.83	10.37
'Rosaline'	33.67	31.33	21.67	10.33	11.00	45.00	52.67	61.00	1.99	11.63
C.D. (5 %)	2.00	2.17	1.58	1.68	1.85	1.64	1.95	1.72	0.10	0.53

On the basis of the investigation, it was concluded that all the seven varieties had variation in respect of growth and flowering under naturally ventilated polyhouse under standard growth and climatic conditions. Three varieties Rosaline, Dana Ellen and Silvester were recommended for commercial cultivation of gerbera for the production of cut-flowers in and around Lucknow, Uttar Pradesh.

References

Barooah, L., Choudhury, Talukdar, M. 2009. Evaluation of different gerbera (*Gerbera jamessonii* Bolus ex Hooker F.) cultivars under agro climatic conditions of Jorhat, Assam. *J. Orna. Horti.,* 12 (2): 106-110.

Barreto, M.S. and Jagtap, K.B. 2006. Assessment of substrates for economical production of gerbera (*Gerbera jamesonii* Bolus ex Hooker F.) flowers under protected cultivation. *J. Orna. Hort.,* 9: 136-38.

Biradur, M.S and Khan, M.M. 1996. Performance of exotic gerbera varieties under low cost plastic green house. *Lal Baugh,* 41(3 and 4):46-52.

Kumar, R. and Yadav, D.S. 2005. Evaluation of gerbera (*Gerbera jamesonii* Bolus ex Hooker F.) cultivars under sub-tropical hills of Meghalaya. *J. Orna. Hort.,* 8: 212-15.

Praneetha, S. 2006. Performance of gerbera (*Gerbera jamesonii* Bolus ex Hooker F.) genotypes at Shervaroy hills of Tamil Nadu. *J. Orna. Hort.,* 9(1): 55-57.

Singh, K.P and Ramachandran, N. 2002. Comparison of greenhouses having natural ventilation and fan pad evaporative cooling systems for gerbera production. *J. Orna. Horti.,* 5(2): 15-19.

Thangam, M., Ladaniya, M.S. and Korikanthimath, V.S. 2009. Performance of gerbera varieties in different growing media under coastal humid conditions of Goa. *Indian J. Hort.,* 66(1):79-82.

Chapter 20

Molecular Characterization of Various Turfgrass Genotypes by RAPD and ISSR Markers

Roshni Agnihotri[1]*, S.L. Chawla*[2] *and Sudha Patil*[2]

[1]*Dr. Y. S. Parmar University of Horticulture and Forestry, Solan (H.P.)*
[2]*ASPEE College of Horticulture and Forestry, Navsari Agricultural University, Navsari, Gujarat*
E-mail: roshniagnihotri134@gmail.com, shivlalchawla@yahoo.com

Abstract

The molecular characterization was performed in thirteen warm season turfgrasses at Department of Plant Molecular Biology and Biotechnology, Navsari Agricultural University, Navsari, Gujarat to establish the genetic affinity/divergence among the genotypes studied for their bio-agronomic characteristics. The tested genotypes included thirteen turfgrass germplasm. Molecular characterization of all the genotypes was performed by RAPD and ISSR molecular markers. Out of 150 decamer random primers used for initial screening, 6 primers were used for characterization. 6 primers yielded a total of 80 scorable bands, of which all of them were polymorphic. The highest genetic similarity coefficient (0.395) was observed between C. dactylon L. 'Bargusto' and C. dactylon L. 'Palma', while, lowest genetic similarity (0.029) was recorded between C. dactylon L. 'Panama' and C. dactylon L. × C. transvaalensis 'Tifdwarf'. In ISSR, a total of 17 primers were screened, of which 6 gave clear amplification pattern. A total of 76 scorable bands in size ranging from 150bp to 1400bp and showed 100 per cent polymorphism. Highest Jaccard's genetic co-efficient (0.750) was observed between C. dactylon L. 'Bargusto' and C. dactylon L. 'Panama', while, lowest genetic similarity co-efficient (0.085) was recorded between C. dactylon L. 'Local' and S. secundatum.

Keywords: *Turfgrass, Molecular markers, RAPD, ISSR.*

Introduction

Understanding the genetic similarity of frequently used germplasm is vital to any breeding program attempting to increase the genetic diversity of new cultivars. An accurate knowledge of the origin and parentage of parental germplasm may also lead to a better understanding of the inheritance of important genetic traits. Genetic markers are a basic tool plant breeders use for cultivar identification, pedigree analysis, and assessing genetic diversity. DNA molecular markers can be used to evaluate genetic variation within and among plant taxa, which have gained wide spread popularity (Paterson *et al.*, 1991; Yang *et al.*, 1996). The PCR technology has offered new marker systems for diagnosis of genetic diversity in large-scale studies (Saiki, *et al.*, 1988). Over the last 25 years, PCR technology has led-to extensive use of of two simple and quick techniques called Random Amplified Polymorphic DNA (RAPD) and Inter Simple Sequence Repeats (ISSR). The development of PCR-based marker systems, especially Randomly Amplified Polymorphic DNA (RAPD) markers (Williams *et al.*, 1990), has been a boon to plant breeders and geneticists. These markers have been used both for DNA fingerprinting (Martín and SánchezYélamo, 2000; Blair, *et al.*, 1999) and population genetic studies (Wolfe, *et al.*, 1998).

Grasses are one of the most complex biological and genetic systems in nature, while gramineae family is defined as a single genetic system which evolved approximately 55-70 million years ago (Kellog, 2000). In many crops, genetic information on diversity can be obtained by pedigree analysis but in tropical turfgrasses this is problematic because in India many cultivars were imported in the late 20th centuries by nurserymen with inadequate record keeping and protection against cross pollination, resulting in many cultivars where only the maternal parent or neither parent has been established. Distinguishing turfgrass cultivars solely upon morphological traits is difficult and has proven challenging for breeders and inspectors charged with certifying cultivar purity (Chen *et al.*, 2009). Several DNA marker techniques have been used to assess nucleotide sequence variation in different grasses. Though there have been many years and levels of investigation, the exact composition, organization and evolution of its genome is still unknown (Buddak *et al.*, 2004). Domestication of turfgrass is recent compared to any other grasses, and is classied into more than 30 species belonging to 20 genera and 3 subfamilies of Gramineae (Gould and Shaw, 1983). In molecular improvement of turfgrass species, there is still a knowledge gap compared to the agronomically important cereal crops (Gresshoff *et al.*, 1998). Therefore, it is important to close this gap and accurately identify turfgrasses through molecular approaches that can answer a broad range of genetic, evolutionary relationships and ecological questions (Huff, 1998).

Genotype Notation Used in Molecular Characterization

Notation	Genotype
T_1	*Cynodon dactylon* L. 'Black Jack'
T_2	*C. dactylon* L. 'Bargusto'
T_3	*C. dactylon* L. 'Panama'
T_4	*C. dactylon* L. 'Palma'
T_5	*C. dactylon* L. 'Panam'
T_6	*C. dactylon* L. 'Selection 1'
T_7	*Axonopus compressus* Beauv.
T_8	*C. dactylon* L. 'Local'
T_9	*Zoysia tenuifolia* Willd. Ex Trin.
T_{10}	*Zoysia matrella* (L.) Merr. Philipp.
T_{11}	*Stenotaphrum secundatum* [Waltz] Kuntze.
T_{12}	*Eremochola ophiuroides* [Munro] Hack.
T_{13}	*Cynodon dactylon* x *Cynodon tranvaalensis* 'Tifdwarf'

Plant Material

In the present study, one known hybrid (*Cynodon dactylon* × *Cynodon tranvaalensis* 'Tifdwarf'), one local selection of unknown parental description (*C. dactylon* L. 'Local'), one commercially exploited and widely distributed variety of India (*C. dactylon* L. 'Selection 1'), five exotic genotypes that are commercial in other countries (*C. dactylon* L. *viz.*, 'Black Jack', 'Panama', 'Palma', 'Panam', 'Bargusto'), two species of genus *Zoysia* (*Z. matrella* and *Z. tenuifolia*) and one genotype each of three different genus (*Axonopus compressus*, *Stenotaphrum secundatum* and *Eremochola ophiuroides*) were characterized.

Methods

DNA Extraction

Genomic DNA was isolated from fresh leaves by using a modifiedCetylTrimethyl Ammonium Bromide (CTAB) extraction procedure (Murray and Thompson, 1980). Fresh leaves were ground to fine powder in liquid nitrogen in a pre-chilled and autoclaved mortar and pestle with a pinch of PolyVinylPyrrolidone (PVP) powder added in it. The powder was then transferred to a 2 ml tube containing 0.5 ml of preheated (65°C) 2X CTAB extraction buffer. The CTAB buffer contained 100 mMTris-HCl, pH 8; 40 mM EDTA stock 0.5 M; 1.5 M NaCl stock, 5 M; ß-Mercapto ethanol 1 per cent; CTAB (CetylTrimethyl Ammonium Bromide) 3 per cent; PVP (Polyvinyl pyrrolidone) 1 per cent and sterile millipore water to make up volumeupto 50 ml. The suspension was incubated at 60°C for 1 hr with intermittent shaking and swirling at every 15 minutes. Equal volume of chloroform: isoamyl alcohol (24:1 v/v) was added to the suspension and content were mixed by inversion to form an emulsion.

The mixture was centrifuged at 10,000 rpm for 10 min at RT (room temperature) so as to separate the phases. The supernatant containing upper phase was transferred to a new sterilized tube such that it avoids shearing and above step was repeated. The aqueous layer was transferred to a new tube, and precipitation was done with 0.6 volume of ice-cold isopropanol and tubes were incubated at -20°C overnight. The precipitated sample was then centrifuged at 10,000 rpm for 25 min at 4°C. The pellet sedimented by centrifugation was washed carefully twice with 70 per cent ethanol, dried at room temperature and re-suspended in 100µl of TE buffer (10mM Tris-Hcl, 1mM EDTA pH-8). The quality and quantity of extracted DNA were determined both by spectrophotometric analysis and gel electrophoresis. Depending on leaf samples, yield in DNA varied from 300ng to 1000ng using this protocol.

Random Amplified Polymorphic DNA Reaction System

Amplification reaction was performed based on the protocols of Williams *et al.* (1990) using decanucleotide primers from Operon Technologies (California, USA) and plant RAPD primer set from GeNei^{TM} (Merck Specialities Private Limited, India). Primary screening of 95 primers were doneand the primers that gave clear and polymorphic amplification patterns were used for further analysis. For each primer, amplification was carried out in a 200 µl thin walled PCR tube containing 20 µl reaction mix volume containing DNAseRNAse free water (Genei) 12.5µl, 10X Taq Buffer (Genei) 2.0 µl, dNTP pre-mix (10 mM) (Biogene) 0.4 µl, 1.5mM $MgCl_2$ (25 mM) (Fermentas) 1.2 µl, Taq polymerase (3 U/µl) (Genei) 0.3µl, DMSO 0.3, BSA (10mg/ml) (Fermentas) 0.3 µl, Primer (10 pmoles/ml) 1.0µl, Genomic DNA (50-60ng) 2.0µl. Polymerase chain reaction was carried out using athermal cycler (Eppendorf and Biometra) for 5 minutes for initial denaturation at 94°C, followed by 40 cycles of 1 minute at 94°C, 1 minute at 38°C and 2 minute at 72°C. The last cycle was followed by 8 minute extension at 72°C and then held at 4°C.

Inter Simple Sequence Repeat

ISSR analysis was carried out with 25 ISSR primers from UBC (University of Columbia).After initial screening of these 25 primers, the primers which gave clear and polymorphic pattern were selected for further analysis. The PCR reactions for ISSR were carried out in a 25µl reaction mixture containing sterile 18.7 µl Millipore water, 2.5 µl 10X Taq Buffer with 1.5mM $MgCl_2$ (Genei), 0.5 µl dNTP pre-mix (10 mM) (Biogene), 0.3 µl Taq polymerase (3 U/µl) (Genei), 1.0 µl Primer (10 pmoles/ml) and 2.0 µl Genomic DNA (50-60 ng) and PCR were performed in a thermal cycler (Biometra and Eppendorf) using a thermal cycling regime of initial denaturation for 5 minutes at 95 °C, followed by 35 cycles of 1 minute at 94°C, 45 second at 38°C and 1.5 minute at 72°C. The last cycle was followed by 8 minute extension at 72°C and then held at 4°C.

Both of the PCR products were analyzed on 1.5 per cent (w/v) agarose gel in 1X TBE buffer containing ethidium bromide. The electrophoresis was carried out at 80-100 V to separate the amplified bands. Along with the samples the DNA ladder of 500bp and 100bp which is ready to use was also loaded. A potential difference of 7-8V/cm was provided till the bands resolved properly.

Scoring and Data Analysis

The RAPD and ISSR profiling were scored for the presence (1) or absence (0) of bands of various molecular weight sizes in the form of binary matrix. Data were analyzed to obtain Jaccard's similarity coefficients among the isolates by using NTSYS-pc (version 2.21a; Exeter Biological Software, Setauket, NY). The SIMQUAL program was used to calculate the Jaccard's similarity coefficients. A common estimator of genetic identity was calculated as follows:

Jaccard's similarity coefficients = NAB/(NAB+ NA+ NB)

where, NAB is the number of bands shared by samples, NA represents amplified fragments in sample A, and NB represents fragments in sample B. Similarity matrices based on these indices were obtained which were utilized to construct the UPGMA (unweighted pair-group method with arithmetic average) dendrograms. All the gels were scored twice manually and independently. All unique bands were also scored and included in the analysis. Presence or absence of unique, shared, and polymorphic bands was used to generate similarity coefficients. The Jacquard's similarity coefficient was calculated by formula:

$$\text{Similarity coefficient} = \frac{\text{No. of polymorphic band}}{\text{Total no. of bands}}$$

Principal Coordinate Analysis (PCO) was conducted by using module DCENTER and EIGEN of NTSYSpc to complement the cluster analysis.

Result

The primary screening DNA amplification using the PCR for RAPD analyses involving 150 primers resulted in an average of 5.4 (range 1–12) discernible DNA fragments per oligonucleotide primer-DNA template combination. The quality of DNA amplification and readability of DNA bands in agarose gels varied considerably among primers. Many of the primers (≈ 50 per cent) produced either no amplification or generally unreadable amplification patterns (DNA "smear") and in few some genotypes showed no bands, thus could not be scored for RAPDs. Innis and Gelfand (1990) have suggested that such problems (particularly DNA smears) maybe a result of nonspecific amplification during initial PCR cycles and subsequent preferential amplification of nonspecific products during later PCR cycles ("plateau effect"). Williams *et al.* (1990) reported that smears could be minimized by reducing either the polymerase or the genomic DNA in reactions. Out of 150 decamer random primers used for initial screening of 13 genotypes, 77 primers gave no amplification at all, while 6 amplified polymorphic patterns. These 6 primers were then used for RAPD analysis of all 13 genotypes. Amplification products of 13 genotypes with these 6 primers yielded a total of 80 polymorphic scorable bands. The size of amplification products ranged from 100 bp to 1.5 kb. The highest number of bands (23) was obtained with primer OPA-18, while lowest number of bands (10) was obtained with primer OPA-19, OPN-6 and RPI-12. All the 6 RAPD primers used for the present analysis showed 100 per cent polymorphism. The RAPD amplification data were used to obtain a similarity matrix and for generation of dendrogram (Figure 20.1) using UPGMA method. Cluster A Jaccard's Genetic Co-efficient shows the high level of genetic variation among 13 turfgrass genotypes which ranged between 0.029 and 0.395. The highest genetic similarity coefficient 0.395 was observed between T_2 and T_4, while the lowest genetic similarity coefficient 0.029 was between T_3 and T_{13}. It has an average of 0.212 genetic similarities.

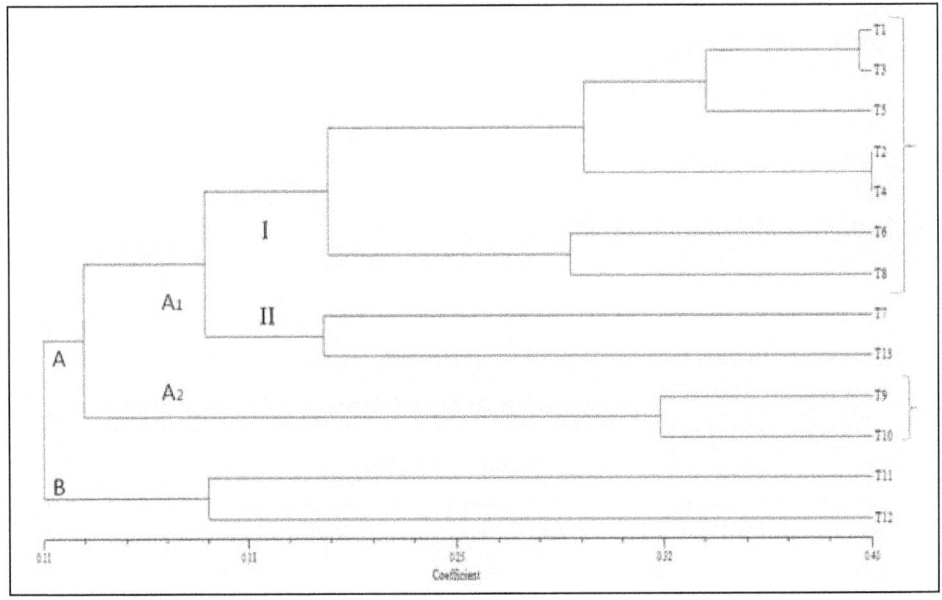

Figure 20.1

A dendrogram was generated (Figure 20.1) which signified the overall genetic relationship among the turfgrass cultivars. The dendogram clearly indicated two main clusters A and B. According to the analysis, 13 genotypes of turfgrass were grouped in two main clusters at coefficient level of 0.12. Cluster A and B showed a similarity of 0.110. Cluster A was bifurcated in to two sub-clusters, A1 and A2 with similarity coefficient of 0.122. Sub-cluster A1 again grouped in to two group I and II in which group I contained genotypes, T_1, T_2, T_3, T_4, T_5, T_6 and T_8 which represented the varieties belonging to *Cynodondactylon* L. and group II contained genotypes T_7 and T_{13} which showed similarity index of 0.207. Cluster A2 consisted of genotype T_9 and T_{10} which belonged to genus *Zoysia* and showed similarity index of 0.323. Cluster B contained T_{11} and T_{12} depicted a similarity index of 0.167.

ISSR markers

A total of 17 Primers consisted of di and tetra nucleotide repeat motifs were used for initial screening. Out of these, 5 primers gave no amplification at all, while only 6 primers were found to give clear banding patterns and were subsequently used to analyze the entire set of 13 genotypes.Out of these 6 primers, UBC-810 contained (GA)n repeat motif, UBC-821 contained (GT)n motif, UBC-825contained (AC)n motif, UBC-857 contained (AC)nYG motif, UBC-873 contained (GACA)n motif and UBC-875 contained (CTAG)n motif. These 6 ISSR primers amplified a total of 76 scorable bands in size ranging from 150bp to 1400bp of which 76 bands were polymorphic thus they showed 100 per cent polymorphic banding pattern.

All the 6 ISSR primers used for the present analysis showed 100 per cent polymorphism. UBC-821 generated a minimum of 7 bands and the maximum of 17 bands were generated by UBC-873. The ISSR amplification data were used to obtain a similarity matrix.Jaccard's Genetic Co-efficient shows the high level of genetic variation among 13 turfgrass genotypes which range between 0.085 and 0.750. The highest genetic similarity coefficient 0.750 was observed between T_2 and T_3, while the lowest genetic similarity coefficient 0.085 was between T_8 and T_{11}. It has an average of 0.417 genetic similarities. A dendrogram (Figure 20.2) was generated using UPGMA method, which signifies the overall genetic relationship among the turfgrass cultivars.The dendrogram clearly indicated two main clusters A and B. According to the analysis, 13 genotypes of turfgrass were grouped in two main clusters at coefficient level of 0.34.

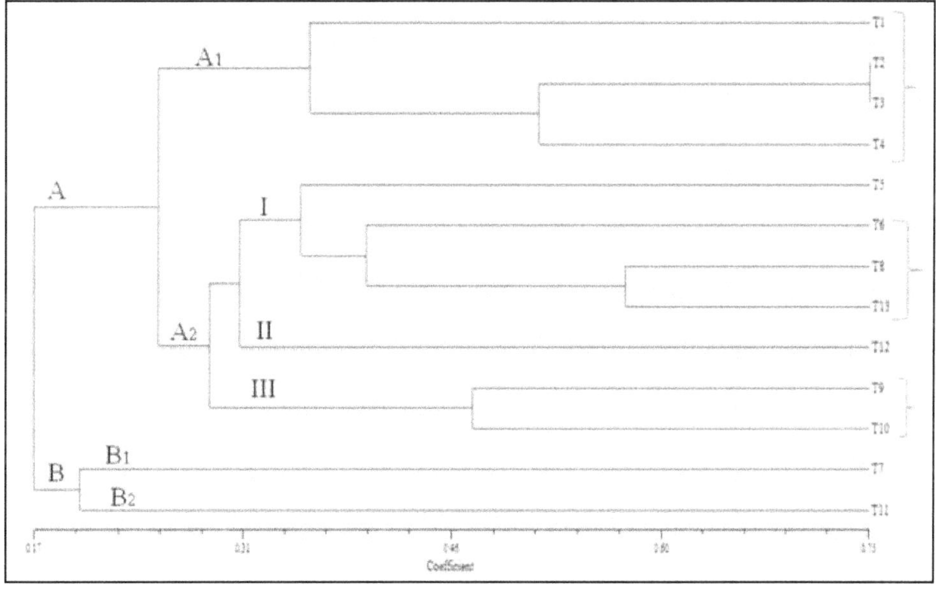

Figure 20.2

Cluster A and B showed a similarity of 0.170. Cluster A was bifurcated in to two sub-clusters, A1 and A2 with similarity coefficient of 0.262. Sub-cluster A1 consisted genotypes T_1, T_2, T_3 and T_4, which belonged to same species *Cynodon dactylon*L., while, sub-cluster A2 again grouped in to three group I, II and III in which group I contains genotype T_5, T_6, T_8 and T_{13}, which consisted of three small leaved dwarf genotypes of *Cynodon* genus, while one morphologically tall growing genotype was also in the same group, although, they showed high dissimilarity coefficient among themselves. Genotype 12 grouped in to distinct cluster group II as this belonged to a morphologically and taxonomically distinct genus *Eremocholao phiuroides*. Group

III contains genotype T_9 and T_{10}, which are members of genus *Zoysia* with similar growth habit and morphology.

The cluster B was further divided in to two sub-clusters B1 and B2 with similarity coefficient of 0.211. Sub-cluster B1 consisted of genotype T_7 which was a distinct broad leaved genus of poaceae family *viz. Axonopus compressus,* while sub-cluster B2 consisted of genotype T_{11} which also belonged to a distinct taxonomic genus *Stenotaphrum,* which had wider leaf blades and waxy leaves. The amplification data obtained from RAPD and ISSR primers were used for similarity matrix and generated combined dendrogram depicted in Figure 20.3.

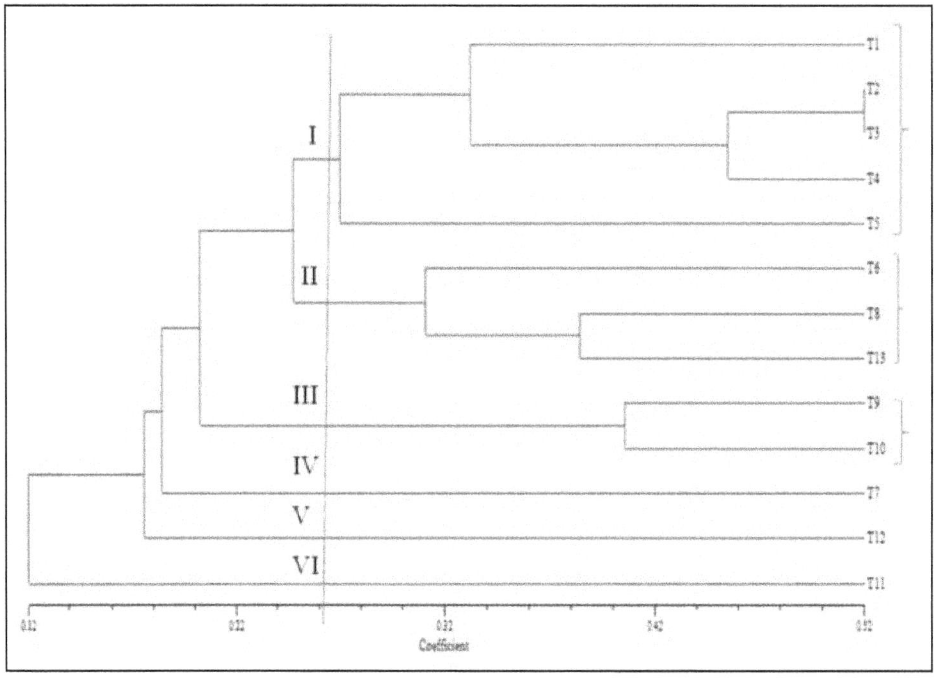

Figure 20.3

Jaccard's Genetic Co-efficient showed the high level of genetic variation among 13 turfgrass genotypes ranging between 0.067 and 0.520. The highest genetic similarity coefficient (0.520) was observed between T_2 and T_3, while the lowest genetic similarity coefficient (0.067) was observed between T_8 and $T_{11.}$ It had an average of 0.293 genetic similarities. A dendrogram was generated, which signified the overall genetic relationship among the turfgrass cultivars and corresponded well with dendrogram of ISSR.

Principle coordinate analysis (PCA) based on genetic similarity matrices were used to visualize the genetic relationship of 13 genotypes of turfgrass (Figure 20.4). Only polymorphic bands were used to construct a binary matrix. PCA performed to visualize the dispersion of genotypes in relation to the two principle axes of variation. It is clear from plot (Figure 20.4) that the genotypes were relatively spread out on both axes. The genotypes T_1, T_2, T_3, T_4 and T_5 were grouped together because they belonged to *Cynodon dactylon* L. and showed similar morphological behaviour. While, genotypes T_9 and T_{10} were grouped into other group as they belonged to genus *Zoysia*. T_{13}, T_8 and T_6 formed a distinct cluster, as corresponding to their morphological similarity of being dwarf, slow growing and small leaved genus of *Cynodon*. However, rest wider leaved species (T_7, T_{11}, T_{12}) were randomly allocated on the 3-D coordinate. Cluster I of combined dendrogram of RAPD and ISSR grouped the exotic *Cynodon* germplasm into one cluster with similarity coefficient of 0.27 owing to their origin from temperate region. Contrary to expectations, all exotic genotypes of *Cynodon* did not form a single cluster in individual dendrograms of RAPD and ISSR and displayed a wide range of variation, and this corresponds to their different regions of origin.

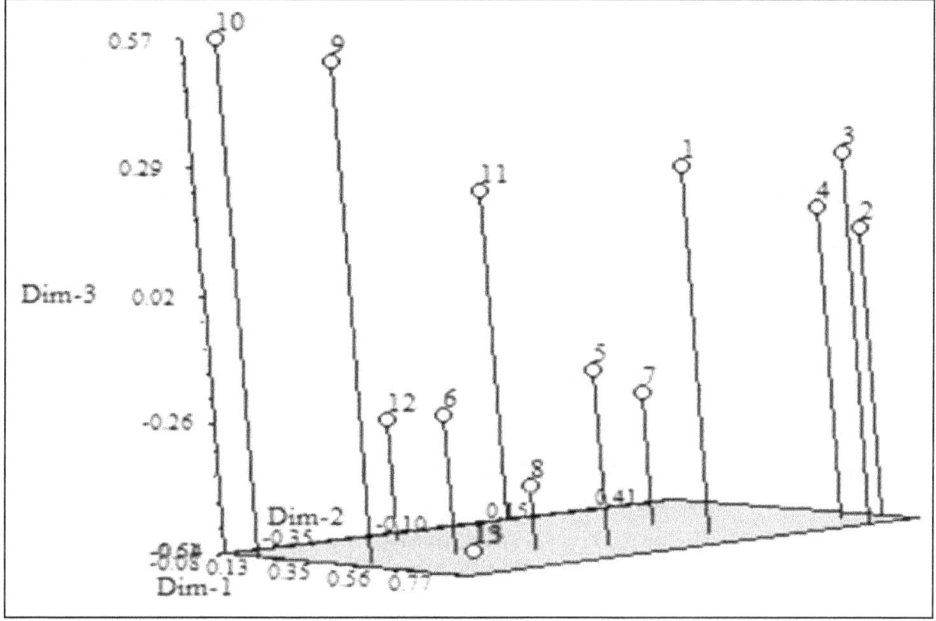

Figure 20.4

Conclusion

The results obtained in this study demonstrated that RAPD and ISSR analysis together could be used effectively to detect genetic diversity among turfgrass genotypes with different origins. In the same study a good correlation was found between morphological characteristics and clustering pattern resulted from molecular analysis.

References

Blair M.W., Panaud O. and Mccouch, S.R. 1999. Intersimple sequence repeat (ISSR) amplification for analysis of microsatellite motif frequency and fingerprinting in Rice (*Oryza sativa* L.). *Theor. Appl. Genet.* 98:780-792.

Budak H., Shearman R.C., Gulsen O. andDweikat I. 2005. Understanding ploidy complex and geographic origin of the *Buchloedactyloides* genome using cytoplasmic and nuclear marker systems. *Theor. Appl. Genet.* 111(8):1545-1552

Chen Z., Wang, M.L., Waltz, C. andRaymerl, P. (2009). Genetic diversity of warm-season turfgrass: seashore paspalum, bermudagrass and zoysiagrass revealed by AFLPs. *Floriculture and Ornamental Biotechnology.* 3(1): 20-24.

Gresshoff P.M., GallahanL.M., Ghassemi F. and Caetano-Anolles G. 1998. Molecular genetic analysis of turfgrass, p. 3–18. In: M.B. Sticklen and M.P. Kenna (eds.). Turfgrass biotechnology: Cell and molecular genetic approaches to turfgrasss improvement. Ann Arbor Press, Chelsa, Mich.

Gould F.W. and Shaw R.B. 1983. Grass systematic. Texas A and M Univ. Press, Dallas.

Huff D.R. 1998. Genetic characterization of open-pollinated turfgrass cultivars, p. 19–30. In: M.B. Sticklen and M.P. Kenna (eds.). Turfgrass biotechnology: Cell and molecular genetic approaches to turfgrasss improvement. Ann Arbor Press, Chelsa, Mich.

Innis M.A. and Gelfand D.H. 1990. Optimization of PCRs, p. 3–12. In: M.A. Innis, D.H. Gelfand, J.J. Sninsky, and T.J. White (eds.). PCR protocols—A guide to methods and applications. Academic Press, New York.

Kellog E.A. 2000. Evolutionary history of the grasses. *Plant physiology.* 1205: 1198-1205.

Martín J.P. and Sánchez-Yélamo M.D. 2000. Genetic relationships among species of the genus *Diplotaxis* (Brassicaceae) using inter-simple sequence repeat markers. *Theor. Appl. Genet.* 101:1234–1241.

Murray M. G. and Thompson W. F. 1980. Rapid isolation of high molecular weight plant DNA. *Nucleic Acid Res.* 8:4321-4325.

Paterson A.H.,TanksleyS.D. and SorrellsM.E. 1991. DNA markers in plant improvement. *Adv. Agron.* 46:39–90.

Saiki R.K.,Gelfond D.H., Stoffel S., Scharf S.J. and Higuchi R. 1988. Primer-directed enzymatic amplification of DNA with a termostable DNA polymerase. *Science.* 239:487–491.

Williams J.G.K., Kubelik A.R., Livak K.J., RafalskiJ.A. and Tingey S.V. 1990. DNA Polymorphisms amplified by arbitrary primers and useful as genetic markers. *Nucl. Acids Res.* 18: 6531-6535.

Wolfe A.D., Xiang Q.Y. and Kephart S.R. 1998. Assessing hybridization in natural populations of *Penstemon* (Scrophulariaceae) using hypervariable intersimple sequence repeat (ISSR) bands. *Mol. Ecol.* 7:1107-1125.

Yang W.P., Oliveira A.C., Godwin I., Schertz K. andBennetzen J.L. 1996. Comparison of DNA marker technologies in characterizing plant genome diversity: variability in Chinese sorghums. *Crop Sci.* 36:1669–1676.

Chapter 21

Seasonal Flowering Plants for Healthy Environment in Landscape Gardening

Shankar Verma and Rajeev Kumar

CSIR-National Botanical Research Institute, Rana Pratap Marg, Lucknow (U.P.)

Abstract

In floriculture, seasonal flowering plants play an important role in developing beautiful healthy environment. Present subject deals with the season bound flowering plants which commonly termed as seasonal annuals. Flowering seasonal plants were assessed as winter season, summer season and rainy season annuals. Seasonal annuals arranged in beds according to their height they were marked as for growth and development in first line of beds. Behind the first line of dwarf annuals, medium height annuals seedlings are raised and planted for medium height. Tall annuals are transplanted in back third/fourth lines of beds. Such plantings were developed with main idea of colour arrangement which were found most beautiful. Varying shades of colours of Pansy, Calendula, Brachycome, Aster, Sweet alyssum, Gaillardia, Nemesia, Coriopsis, Daisy, Gazania, Gypsophila, Linaria, Mesembryanthemum, Mimulus, Nusturtium, Phlox, Poppy, Salvia, Cineraria, Verbena are suitable for first line at edge of the suitable lines. In second line behind the first line beds, the suitable plants were found namely - Acroclinum, Arctotis, Antirrhinum, Clianthus, Companula, Delphinium, Dianthus (Sweet William), Carnation, Gazania, Lady's Lace, Marigold, Nicotiana, Poppy, Saponaria, Stock, Mathiola etc. Thus planted tall plants in back line for flowering used to give best effect. Annual Chrysanthemum, Calceolaria, Argimon, Cheiranthes, Clarkia, Dahlia, Delphinium, Godetia, Helichrysum, Hollyhock etc. were used was planting. Some annual creepers also give their best creation in landscaping like Minalobeta, Sweet pea, Specha, Morning glory. Flowering annuals in different seasons have been arranged for plantings.

Keywords: *Season, Winter, Annuals, Flowering. Creepers.*

Introduction

In the early periods ornamental flowering annuals and bulbous plants were preferred for different purposes (Jankiram *et al.*, 2013). Seasonal ornamental flowers are very indispensable for decorative value in the gardens, parks, houses, in any spot for beautification. They are grown in different particular seasons in the world (Bose and Yadav, 1990; Raghava, 1992). In temperate region some of the annuals remain alive for years but produce flower once in the season in a year. In fact annuals flower in a season according to climatic conditions. Therefore flowering of annuals vary with the temperature of the season (Tomar, 1999). However, Misra and Misra, 2008 described them in quarter basis for flowering of Asiatic lilies, *Clematic, Gazania, Gaillardia, Gladiolus, Chrysanthemum etc.* in July-August season. Nursery management is an essential and important work for growing ornamental seasonal flowering plants. Seedlings of seasonal annuals are raised in the nursery. Mixed soil is covered with polythene or gunny bag for 35 to 45 hours, caption or bavistin fungicides may be used to sterilize the soil.

Selection of good quality healthy seeds and their proper sowing are important factors for raising of seasonal annuals. Seed sowing times vary in consideration with the season and climate mainly winter season annuals, summers and rainy season annuals. Sowing of annual herbaceous and perennials minute seeds need special care for proper distribution in bed or pot. These are covered with sieved light leaf mould soil and very light spray watering should be given for providing moisture to germinate the seeds. Seedlings are transplanted in proper laid out beds or pots. Transplanted annuals and seeded plants should be given proper cultural practices for growth and flowering (Bose and Yadav, 1990; Randhawa and Mukhopadhyay, 1959).

Materials and Methods

Seeds of annuals were sown for raising seedlings for about 25 to 40 days before planting. Seeds were sown in raised beds taking consideration for watering plant protection measures in seedlings raising. Healthy plant material of annuals always produces good flowers resulting from the normal growth performance. Nursery soil was clean and healthy, porus, sandy-loam and free from microorganisms. It was essential to eliminate soil borne diseases. Soil was sterilized to save the seedlings from fungus. For sterilizing soil 2.0 per cent formalin solution was mixed with the soil (Panwar, 1999).

Seedlings of Annual flowering plants were planted according to their height and flower colour arrangement in herbaceous border of the experimental plan. All cultural practices were provided to the plants during growing periods and flowering. Selection of plants was done for three groups as per arrangement of 3 lines of beds. (1) Dwarf plants (2) Medium height plants (3) Tall Plants. These are given below for winter season.

I. Tall Plants

Annual Chrysanthemum, *Calceolaria, Argimon, Cheiranthes, Clarkia, Dahlia, Delphinium, Godetia, Helichrysum* Hollyhock.

II. Medium Plants

Acroclinum, Arctotis, *Antirrhinum, Clianthus, Companula, Delphinium, Dianthus* (Sweet Willium), Carnation, *Gazania, Helichrysum,* Lady's Lace, Marigold, *Nicotiana,* Poppy, *Saponaria,* Stock etc.

III. Dwarf Plants

Pansy, Calendula, Brachycome, Aster, Sweet alyssum, *Gaillardia, Hymenantherum, Nemesia, Coriopsis,* Daisy, *Gazania, Gypsophila, Linaria, Mesembranthemum, Mimulus, Nusturtium, Phlox,* Poppy, *Salvia, Verbena, Cineraria* etc.

Results and Disussion

Details of observations revealed that planting of seasonal flowering annuals was found appreciable when it was arranged according to height. Tall plants in back line provided, blooms upto upper most height attaining to 40 to 75 cm colour of different plants of *Dahlia,* Annual Chrysanthemum Hollyhock, *Helichrysum,* Sweet Pea, *Delphinium etc.* were found the best flowering annuals.

Plants of middle line beds showed very effective spectrum of colour arrangement because of situation of plants balancing the height between Tall and Dwarf plants, Medium height plants of 20 to 40 cm level, at full bloom stage were observed to give a appreciable pleasure at site. Suitable plants-*Acroclynum, Dianthus, Campanula, Antirrhinum, Clianthus, Arctotis, Delphinium, Carnation, Gazania,* Marigold, *Nicotiana, Saponaria,* Stock, Poppy *etc.* were found enriching the colours and colour-blend with the colour of dwarf plants of front line. The arrangements and effect of plants, colour have also been found suitable as per height by Bajpai and Kumar (1998) and Randhawa and Mukhopadhyay (1959).

In front line of beds *Pansy, Calendula, Brachycome, Aster, Sweet Alyssum, Gaillardia, Hymenantherum, Nemesia, coriopsis, Daisy, Gazania, Gypsophila, Linaria, Mesembyanthemum, Mimulus, Nusturtium, Phlox, Poppy, Salvia, Verbena, Cineraria, etc.* were found suitable for growing as dwarf flowering annuals. Flowers revealed the best colour in the morning time. In fact different colours appeared in making sense in vista effect from a certain, distance. Such varying colour effect of different flowers has also been observed as colour flash by different scientists (Singh and Elahi, Sindhu and Singh, 1999).

Summer season annuals, *Zinnia, Kochia, Antirrhinum, Petunia, Marigold, Calceolaria, Helichysum, Nicotiana* and *Portulaca* were observed promising in flowering in forenoon time. In open areas these plants required frequent watering in pots and in beds irrigation should be given as for plants requirement. Field observations were found very critical in different situation of soil and its management.

Flowering plants were found suitable for rainy season are *Ageratum, Collinsia, Cockscomb, Coriopsis, Delphinium, Gaillardia, Zinnia, Lineris, Helichrysum, Gyprophila.* Similar growth and flower production have been discussed by Bajpai and Kumar (1998), Randhawa and Mukhopadhyay (1959) and Singh and Elahi (1999).

References

Bajpai, P.N. and R. Kumar (1998). Indoor gardening and arrangement of Plants. *J. Sci. Hort.* **3:** 41-43.

Bose, T.K. and L.P. Yadav (1990). Commercial flowers. Nayer Prakash, 206, Bidhan Sarani, Calcutta-700006, India

Janakiram, T.K., K.V. Prasad and S. Kumar (2013). Export and Import of flowers: Indian Scenario: *Export Oriented Horticulture. pp*: 26-32.

Kumar, S. (1993). Terrarium gardening in glass. Containers. *Ankur l: PP:* 27-28.

Misra, S. and R.L. Misra (2008). Ornamental Gardening in July-August. *Indian Hort.* 53 (4): 36-37.

Panuar, R.S. (1999). Production of Quality flower of gladiolus. *Delhi-Agri. Hort. Soc. PP*: 9-10.

Raghava, S.P.S. (1996). Seasenal flowers. *Delhi-Agri-Hort. Soc. pp*: 13-14.

Randhawa, G.S. and S. Mukhopadhyay (1959). Floriculture in India, *Indian Council of Agriculture Research,* New Delhi, india.

Sharga, A.N., J.N. Sachan and R.K. Roy (2008). In your garden, Grow new Canna flowering varieties. *Indian Hort.* **53** (4): 35.

Sindhu, S.S. and M.K. Singh (1999). Grow Lilium for Commera. *Delhi Garden. Magzin. PP:* 11-12.

Singh, H. and S. Elahi (1999). Gerbera-A Commercial Flower. *Delhi Grden Magazin. PP*: 7-8.

Singh, J. and S. Elahi (1999). Splash of colours at your doors step. *Ibid. PP:* 21-22.

Tomar, B.S. (1999). Grow marigold, Beyond the winter season. *Delhi Garden Magazine, PP*: 19-20.

Yadav, T.P.S. and V.P. Singh (2008). Cultivation of flowers is a flourishing business in Delhi. *Indian Hort.* 53 (4): 33-34.

Chapter 22

Studies on Vase Life of Cut-Flowers in Dahlia Genotypes

Kiran Singh

Pt. Deen Dayal Upadhyay Govt. Girls Post Graduate College, Rajajipuram, Lucknow, U.P.

Abstract

Present experiments were conducted to enhance vase life of cut flowers of ten varieties of Dahlia. Silver Nitrate and Sucrose solutions were used as holding treatments. $AgNO_3$ and sucrose treatments were found useful for enhancing the vase life of cut flowers. $AgNO_3$ 25ppm + 2.0 per cent sucrose treatment (T_2) was found the best of all other treatments. Vase life of Pompon, Royal Rise, Chicago and Snow Hill Rose was found appreciable for 12.9, 11.0 and 11.5 days in treatment $AgNO_3$ at 25ppm + 2.0 per cent Sucrose, respectively. In overall observations, findings revealed enhanced flower diameter together with flower freshness and vase life.

Keywords: Dalhia, Vase-life, Genotype, Silver nitrate, Sucrose.

Introduction

Post harvest quality and longevity of cut flowers is dependent on several pre and postharvest factors (Halevy and Mayak, 1974, 1981). Among postharvest factors application of various chemical is known for prolonging vase life, increasing the flower size and maintaining the turgidity and colour of flower petals (Deambrogio *et al.*, 1991; Lownds *et al.*, 1994). The vase life of cut flowers is often very short because of water stress symptoms such as wilting and bending of petals below the flower head. Water stress condition reduces the longevity of cut flowers (Srivastava *et al.*, 2005; and Sharma *et al.*, 2005). The reserved food materials in the stem is of utmost importance in accomplishing flower longevity.

Dahlia is an important commercial flower in floriculture crops highly suitable for flower beds, borders and pots. A large number of varieties developed by selection have been used in various purposes of decoration which his handsome income entrepreneurial venture to the corporates. Its different sized cut flowers are marketed for different purposes (Halevy and Mayak, 1974, 1981; Murli, 1990, Murali and Reddy 1993; Roychoudhary and Sarkar, 1995). Several chemicals have been used for prolonging vase life of cut flowers by keeping them in holding solutions (Sang *et al.*, 1996; Sindhu and Pathania, 2003; Yoo and Kim, 2003). Plant growth regulators alonwith sucrose concentrations prolong vase life of cut flowers (Shekh and Johan, 2005; Jabeen *et al.*, 2008). Cut flowers of bulbous plants *i.e.* gladiolus, Lilies, gerbera *etc*. response well to increase size and prolong. Vase life (Lownds *et al.*, 1994; However, Reid (1992) described several factors in prolonging vase life of cut flowers after post harvest technical conditions.

Materials and Methods

The present study was conducted at Chandra Shekhar Azad Agriculture University and Technology, Kanpur during 2011-12 and 2012-13. The vase life of flowers and evaluation of treatments effect for enhancing life with quality of freshness was for investigations. Flowers of Dahlia were harvested from plants at nearly fresh quality opened stage in the morning hours to avoid excessive heat at noon.

The flower stems 15 cm long with woody base were placed randomly in glass bottles containing 750 ml of aqueous solutions of different treatments. The experimental bottles were kept held in the laboratory at about ambient temperature 22°C + 2°C with relative humidity ranging from 65 to 75 per cent.

The experiments were carried out with the treatment of Silver nitrate ($AgNo_3$ 15ppm + sucrose 2.0 per cent) in first set (T_1) and in next set ($AgNo_3$ 25ppm + sucrose 2.0 per cent) (T_2). Cut flowers of Chicago, Royal Rise, Snow Hill Rose, Sun Rise, Pompon, Eternity, Park Beauty, Lord Krishna, Alpana, Black Out. Variety were taken and laid out in Randomized Block Design with three replications. Data recorded on observations were given in Tables 22.1 and 22.2.

Results and Discussion

Results of treatments revealed encouraging response to vase life of cut flowers in dahlia varieties. $AgNo_3$ and sucrose concentrations showed effect best on flower form, its firmness and petal shape. Varietal response to treatments for their performance vase life was found beneficial. Present findings are in accordance with the results obtained in Gladiolus and Gerbera by Murali, (1990) and Jakeen *et al.* (2008), respectively.

Table 22.1: Vase Life of Flowers in Holding Solutions of $AgNO_3$ and Sucrose (2011-12)

Varieties	Treatment Symbol	Holding Solutions + Sucrose	Diameter of Flower (cm)		Vase Life (Days)
			5th day	8th day	
Chicago	T_1	$AgNo_3$ (15pm) + 2.0 per cent Sucrose	22.870	22.150	10.8
	T_2	$AgNo_3$ (25pm) + 2.0 per cent Sucrose	22.750	22.275	11.5
Royal Rise	T_1	$AgNo_3$ (15pm) + 2.0 per cent Sucrose	21.450	21.235	10.7
	T_2	$AgNo_3$ (25pm) + 2.0 per cent Sucrose	21.710	21.875	11.0
Snow Hill Rose	T_1	$AgNo_3$ (15pm) + 2.0 per cent Sucrose	21.350	21.200	10.7
	T_2	$AgNo_3$ (25pm) + 2.0 per cent Sucrose	19.750	20.750	10.5
Sun Rise	T_1	$AgNo_3$ (15pm) + 2.0 per cent Sucrose	18.355	19.725	10.8
	T_2	$AgNo_3$ (25pm) + 2.0 per cent Sucrose	18.410	18.360	10.5
Pompon	T_1	$AgNo_3$ (15pm) + 2.0 per cent Sucrose	15.950	15.900	12.7
	T_2	$AgNo_3$ (25pm) + 2.0 per cent Sucrose	16.410	16.750	12.9
Eternity	T_1	$AgNo_3$ (15pm) + 2.0 per cent Sucrose	23.650	22.750	9.8
	T_2	$AgNo_3$ (25pm) + 2.0 per cent Sucrose	23.400	23.550	9.7
Park Beauty	T_1	$AgNo_3$ (15pm) + 2.0 per cent Sucrose	23.150	23.350	10.5
	T_2	$AgNo_3$ (25pm) + 2.0 per cent Sucrose	23.250	23.415	10.4
Lord Krishna	T_1	$AgNo_3$ (15pm) + 2.0 per cent Sucrose	24.350	23.950	9.2
	T_2	$AgNo_3$ (25pm) + 2.0 per cent Sucrose	24.650	24.360	9.0
Alpana	T_1	$AgNo_3$ (15pm) + 2.0 per cent Sucrose	20.150	21.115	8.2
	T_2	$AgNo_3$ (25pm) + 2.0 per cent Sucrose	20.430	20.655	9.4
Black Out	T_1	$AgNo_3$ (15pm) + 2.0 per cent Sucrose	20.125	20.410	9.6
	T_2	$AgNo_3$ (25pm) + 2.0 per cent Sucrose	19.535	19.640	8.8

Table 22.2: Vase Life of Flowers in Holding Solutions of $AgNO_3$ and Sucrose (2012-13)

Varieties	Treatment Symbol	Holding Solutions + Sucrose	Diameter of Flower (cm)		Vase Life (Days)
			5th day	8th day	
Chicago	T_1	$AgNo_3$ (15pm) + 2.0 per cent Sucrose	22.860	22.145	10.6
	T_2	$AgNo_3$ (25pm) + 2.0 per cent Sucrose	22.730	22.270	11.7
Royal Rise	T_1	$AgNo_3$ (15pm) + 2.0 per cent Sucrose	21.400	21.250	10.5
	T_2	$AgNo_3$ (25pm) + 2.0 per cent Sucrose	21.680	21.732	11.2
Snow Hill Rose	T_1	$AgNo_3$ (15pm) + 2.0 per cent Sucrose	21.300	20.150	10.5
	T_2	$AgNo_3$ (25pm) + 2.0 per cent Sucrose	19.755	20.660	10.6
Sun Rise	T_1	$AgNo_3$ (15pm) + 2.0 per cent Sucrose	18.250	19.670	10.7
	T_2	$AgNo_3$ (25pm) + 2.0 per cent Sucrose	18.400	18.255	10.5
Pompon	T_1	$AgNo_3$ (15pm) + 2.0 per cent Sucrose	15.850	15.640	12.6
	T_2	$AgNo_3$ (25pm) + 2.0 per cent Sucrose	16.350	16.655	12.8

Varieties	Treatment Symbol	Holding Solutions + Sucrose	Diameter of Flower (cm)		Vase Life (Days)
			5th day	8th day	
Eternity	T_1	$AgNo_3$ (15pm) + 2.0 per cent Sucrose	23.650	22.620	9.7
	T_2	$AgNo_3$ (25pm) + 2.0 per cent Sucrose	23.350	23.530	9.6
Park Beauty	T_1	$AgNo_3$ (15pm) + 2.0 per cent Sucrose	23.500	23.100	10.4
	T_2	$AgNo_3$ (25pm) + 2.0 per cent Sucrose	23.240	23.400	10.2
Lord Krishna	T_1	$AgNo_3$ (15pm) + 2.0 per cent Sucrose	24.350	23.750	9.0
	T_2	$AgNo_3$ (25pm) + 2.0 per cent Sucrose	24.750	24.250	9.3
Alpana	T_1	$AgNo_3$ (15pm) + 2.0 per cent Sucrose	20.100	21.110	8.1
	T_2	$AgNo_3$ (25pm) + 2.0 per cent Sucrose	20.415	20.640	9.3

References

Chakraborty, S., P.S. Murli, J. Tarafder and N. Roychoudhury (2008). Effect of Sucrose and trehalose on vase life and flower quality of gerbera. *J. Orna. Hort.* 11 (3): 188-197.

Deambrogio, F., M. Dolci and E. Accati (1991). The effects of keeping solutions on the life of cut *Gerbera* Jamesonii (H. Bolus) Cv. Relecca, flowers. *Adv. Hort. Sci.* 5 (4): 135-138.

Halavy, A.H. and S. Mayak (1974). Improvement in cut flowers quality, opening and longevity by preshimpment treatments. *Acta Horticultural.* 43 335-347.

Halevy, A.H. and S. Mayak (1981). Senexensce and Postharvest Physiology of cut flowers. Part 2. In Horticultural Reviews. Vol. 3 (Janick, J.Ed.) AVI Publishing Westport Conn. pp: 59-143.

Jabeen, A., R. Chandrashekhar and M. Padma (2008). Effect of Plant Growth regulators on Physical Changes and microbial count during vase life of gerbera (*Gerbera jamesonii* Bolus ex. Hook) Cv. *Yanara. J. Orna Hort.* 11 (2): 107-11.

Lownds, N.K., M. Banaras and P.W. Bosland (1994). Postharvest water loss and storage quality of nine pepper (capsicum) cultivars. *Hort. Sci.* 29: 191-193.

Murali, T.P. (1990). Mode of action of metal salts and sucrose in extending the vase life of cut gladiolus. *Acta Horticulturae.* 266: 307-316.

Murali, T.P. and T.V. Reddy (1993). Postharvest life of gladiolus as influenced by sucrose and metal salts. *Acta Horticulturae.* 343: 313-320.

Reid, M.S. (1992). Post harvest Technology of Horticultural crops. *2nd Ed. A.A. Kadr. UCDANR Pub. 3311, pp*: 201-209.

Roychoudhury, N. and S. Sarkar (1995). Influence of chemcials on vase life of gladiolus. *Acta Horticulturae* 405: 389-391.

Sang, C.Y., C.S. Bang, J.S. Lee and D.C. Lee (1996). Effect of Postharvest treatments and preservatives solution on vase life and flower quality of Asiatic hybrid Lily. *Acta Horticulturae.* 414: 277-285.

Sharma, B.P., N.S. Patharia and Y.D. Shrma (2005). Effect of pulsing and storge methods on vase life of Asiatic Lily. *J. Orna. Hort.* 8 (1): 79-80.

Sheikh, M.Q. and A.Q. John (2005). Effect of pulsing chemical combinations on vase life and spike characteristics in gladiolus. *J. Orna. Hort.* 8 (4): 309-311.

Sindhu, S.S. and N.S. Pathania (2003). Effect of Pulsing, holding and low temperature storage on keeping quality of Asiatic Lily hybrids. *Acta Horticulturae.* 624: 389-394.

Srivastava, R., K. Kandpal and S. Jankari (2005). Effect of Polusing solution, Packaging material and storage duration on postharvest life of gladiolus. *J. Orna. Hort.* 8 (2): 115-118.

Yoo, Yong Kweon and Kim Won Sun (2003). Storage solution and temperature affect the vase life of cut gerbera flowers. *Kor. J. Hort. Sci. Tech.* 21 (4): 386-392.

Index

V

W

Z

p.3

Glass House, Lalbagh Botanical Garden. p.8

Bambusetum at JNTBGRI p.9

p.10

Cycad House, CSIR-NBRI, Lucknow. p. 12

Figure 2.1: A Glimpse of IHBT Botanical Garden.

A. Field view, B. A view of Botanical garden, C. *Ginkgo biloba* plantation, D. Ferny, E. *Erythrina blackei*, F. *Prinsepia utilis*, G. *Bauhinia variegata*, H. *Reinwardtia indica*.

p

AJC Bose Indian Botanic Garden
Ministry of Environment, Forest & Climate Change
Botanical Survey of India, Howrah
SCALE

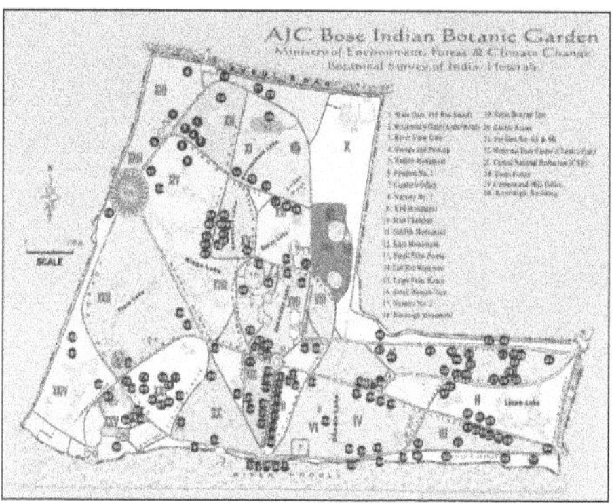

p. 24

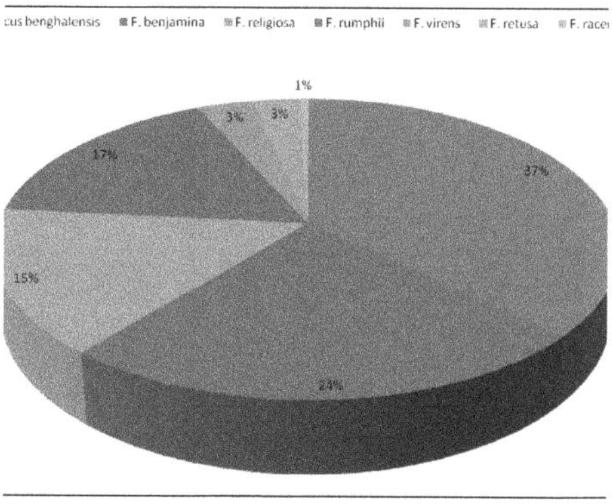

cus benghalensis
F. benjamina
F. religiosa
F. rumphii
F. virens
F. retusa
F. racei
1%
3%
3%
17%
37%
15%
24%

Chart 3.1: Percentage Distribution of different PHE Species. p.41

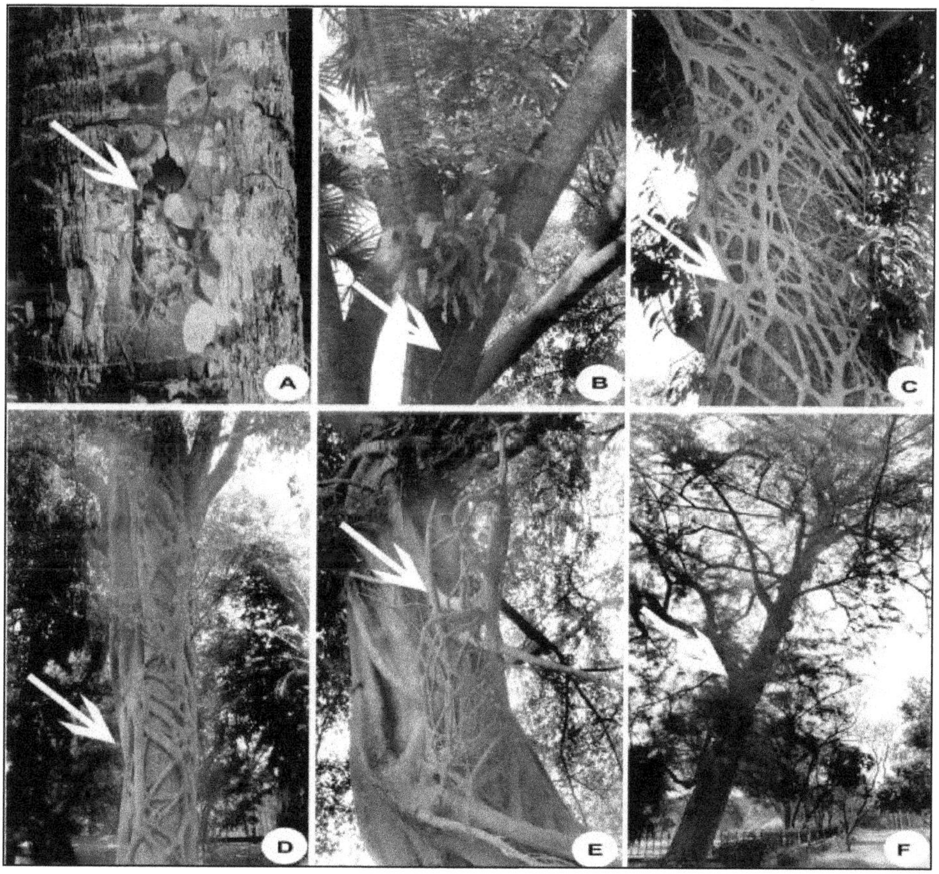

A
B
C
D
E
F

Figure 3.1

A. PHE establishes itself in a crater on the palm trunk, B. Single tap root coming out of the crater, C. Mess of tap root system intermingling together to form a deadly cage around the phorophyte, D. Thickening ofPHE root system, E. Remains of dead phorophyte in a hollow cage, F. Displaced centre of gravity of phorophyte by PHE. p. 42

Figure 3.2

A. *Ficus religiosa* totally covered a *Roystonea regia* plant, B. Dead remains of *R. regia* protruding from the top, C. PHE along with other lianas intermingle a phorophyte, D. Roots of a PHE passing through the dead stem of a unrecognizable phorophyte, E. Cage like structured formed after the death and decay of phorophyte, F. PHE engulfing a concrete structure (Plant Gabion), G. An old building of AJCBIBG affected by PHE. (p.43)

Figure 5.1: Ex-situ Conservation of Few Endemic Plants at National Botanic Garden, Trombay. **(p. 67)**

Figure 5.2: Ex-situ Conservation of Few Rare Plants at National Botanic Garden, Trombay. (p.70)

Figure 5.3: Few Rare Ornamental Trees Conserved at National Botanic Garden, Trombay. (p.73)

Figure 7.1

A. *Catamixis baccharoides* Thomson, B. *Frerea indica* Dalzell, C. *Incarvillea emodi* (Royle ex Lindl.) Chatterjee, D. *Jasminum parkeri* Dunn, E. *Nepenthes khasiana* Hook.f., F. *Phlomoides superba* (Royle ex Benth.) Kamelin and Makhm., G. *Pittosporum eriocarpum* Royle, H. *Sophora mollis* (Royle) Baker. (p.88)

Figure 8.1: Panoramic Satellite View of Botanic Garden of Indian Republic, BSI, MOEF and CC, Govt. of India, NOIDA and Sampling Stations. (p.95)

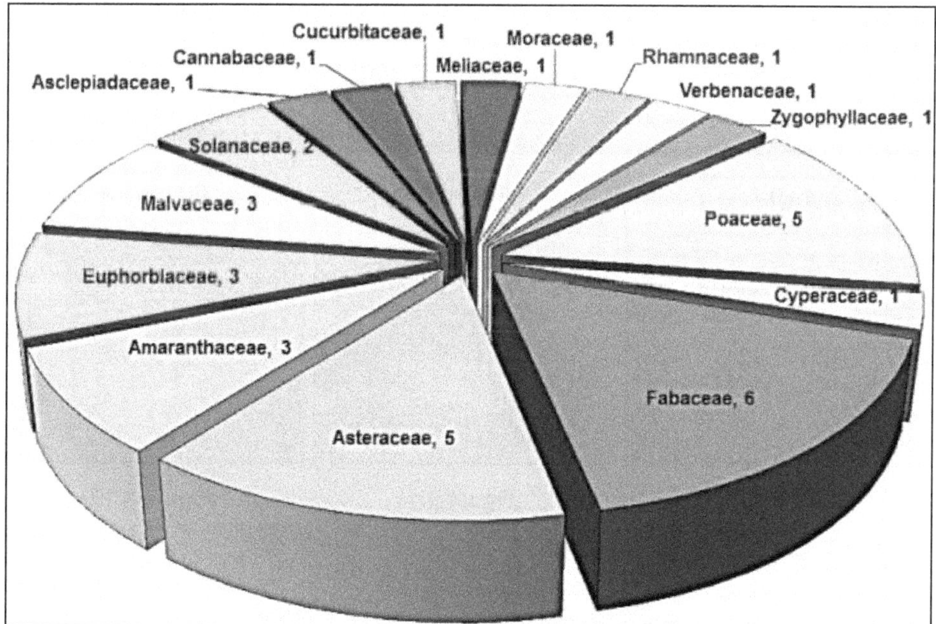

Figure 8.2: Number of Identified Species Belong to different Family of Monocot and Dicot. (p. 98)

Figure 9.1 Figure 9.2 Figure 9.3

Figure 9.4 Figure 9.5 Figure 9.6

Figure 9.7 Figure 9.8 Figure 9.9

Figure 9.10 Figure 9.11 Figure 9.12

Figure 9.13 Figure 9.14 (p.114) Figure 9.15

Figure 11.1: Satellite Image of GNDU Botanical Garden. (p. 129)

Figure 11.3: GNDU Botanical Garden Central Vista. (p.130)

Figure 11.4: ***Sterlizia reginae*** **Aiton.**

Figure 11.5: ***Juniperus chinensis*** **L.**

Figure 11.6: ***Russelia floribunda*** **Kunth.**

Figure 11.7: ***Bauhinia vahlii*** **Wight and Arn.** (p.135)

Figure 11.8: *Ficus infectoria* Roxb.

Figure 11.9: *Casuarina equisitifolia* L. (p.136)

Figure 11.10: *Oroxylum indicum* Vent.

Figure 11.11: *Oncoba spinosa* Forsk.(p. 136)

Figure 14.1: A View of the Canna Garden of National Botanic Garden, Trombay. (p. 163)

Figure 14.2: Some Beautiful Canna Cultivars Conserved at National Botanic Garden, Trombay. (p.164)

Figure 15.1: Bougainvillea Germplasm Collection at CSIR-NBRI. (p.171)

Figure 15.2: Bougainvillea Garden in CSIR-NBRI. (p.171)

Figure 15.3: Dr. P.V. Sane (CSIR-NBRI New Variety-2011). (p.175)

Figure 15.4: Dr. APJ Abdul Kalam (CSIR-NBRI New Variety-2015). (p. 175)

Figure 17.1. Model Polyhouse for Gerbera Cultivation (p.191)

Figure 17.2. Inner view of Gerbera Polyhouse, CSIR-NBRI Botanic Garden, Lucknow. (p. 196)

Figure 19.1. Silvester, Rosaline, Dana Ellen (left to right) (p.209)

www.ingramcontent.com/pod-product-compliance
Ingram Content Group UK Ltd.
Pitfield, Milton Keynes, MK11 3LW, UK
UKHW021532300726
14060UKWH00011B/368

9 789390 3713